RESEARCHED BY
FRANK HAY, JENNI MELDRUM, LEN SAUNDERS, JAMES WALKER

THE BRIGGERS

THE STORY OF THE MEN WHO BUILT THE FORTH BRIDGE

WRITTEN BY ELSPETH WILLS

D1341506

BIRLINN

ARUP

As designers of the second Forth Road Bridge we support the project to build a memorial in memory of the Briggers

Balfour Beatty

100 YEARS 1909-2009 **BUILDING THE FUTURE**

Supported by Balfour Beatty Civil Engineering, main contractor for the restoration of the bridge 2002–2012

First published in 2009 by

BIRLINN LIMITED
West Newington House
10 Newington Road
Edinburgh EH9 1QS

www.birlinn.co.uk

Text copyright © The Briggers Partnership 2009

The moral right of Elspeth Wills to be identified as the author of this work has been asserted by her in accordance with the Copyright, Designs and Patents Act 1988

ISBN: 978 1 84158 761 5

British Library Cataloguing-in-Publication Data
A catalogue record for this book is available from the British Library

Designed and layout by Mark Blackadder

Printed and bound by Bell & Bain Ltd, Glasgow

Previous page.
Briggers on Queensferry cantilever

THE
BRIGGERS

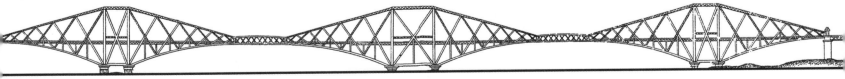

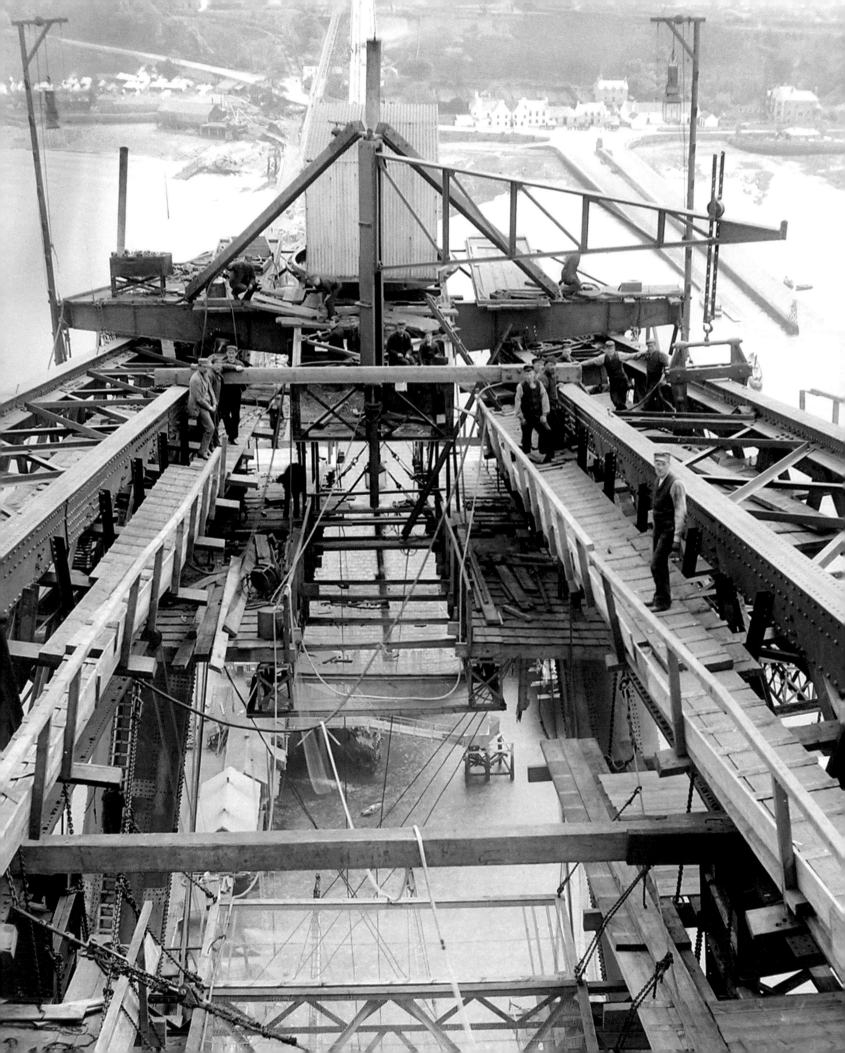

*This book is dedicated
to the Briggers who lost their
lives during the construction of
the Forth Bridge and to their
families and descendants*

CONTENTS

Opposite.
Team using hydraulic riveting machine

PREFACE

Since its opening in 1890 the Forth Bridge has been recognised as one of the world's greatest engineering feats. The human story, however, of the thousands of men who built it – the Briggers – has gone untold.

One man did not forget the Briggers. Growing up within its shadow, Jim Walker has explored every aspect of the history of the bridge over the years. Always at the heart of his research has been his desire to recognise the Briggers, especially those who lost their lives. Jim's enthusiasm proved infectious and local historians Jenni Meldrum and Len Saunders joined his dedicated quest to discover more about the men. Their research was given added urgency by the proposal to erect a memorial to the 57 men believed to have died. A seemingly simple task of identifying them turned out to be a fascinating, challenging and often frustrating piece of detective work. Years of research uncovered only twelve names.

A breakthrough came when Frank Hay joined the team as he was able to enlist the help of two genealogists. The list of deaths grew to 73. With names, the Briggers came alive as individuals. The team wanted to find out more about how they lived and worked and the impact this temporary invasion had on the communities on both sides of the Forth. The idea of writing a book gradually emerged from the mass of evidence acquired. With modern digital photographic techniques, for the first time the faces of individual Briggers could be seen. The men emerged from the shadows of their bridge.

The story does not end with the publication of this book. Hopefully it will not only give the Briggers the recognition that is long overdue but will prompt memories and reveal new sources.

Frank Hay, Jenni Meldrum, Len Saunders, Jim Walker
Queensferry, Scotland

Opposite.
Assembling cantilever tube at drill roads

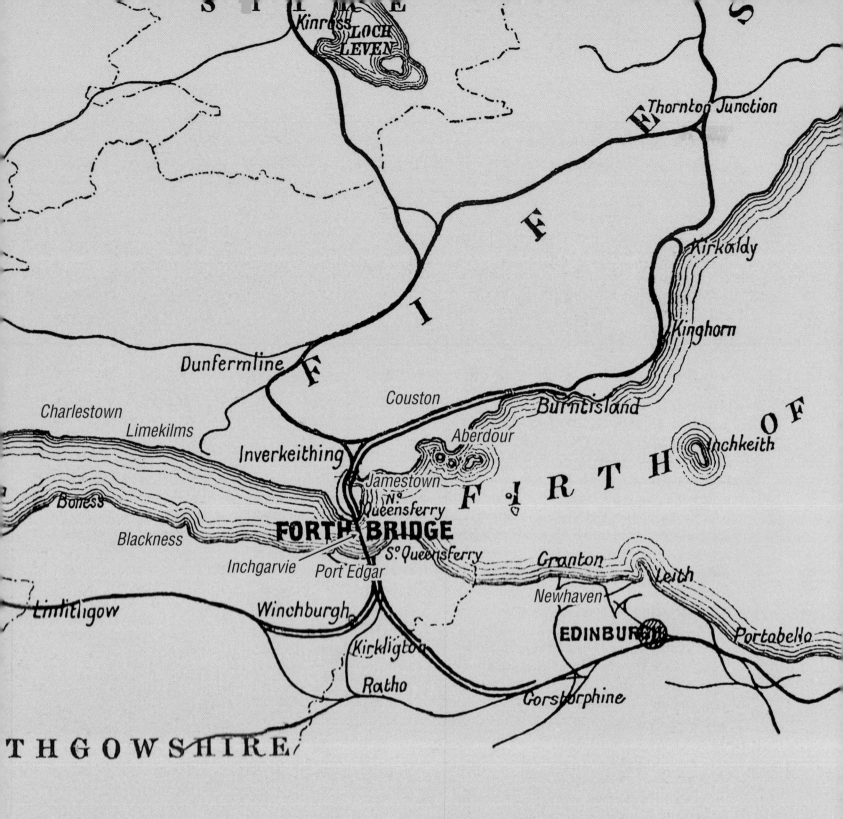

PLAN
showing the
FORTH BRIDGE,
— AND ITS —
RAILWAY CONNECTIONS.

CHAPTER 1
WATER, STEAM AND STEEL

For centuries people must have looked longingly across the mile-long stretch of water that marks the narrowest point of the Forth estuary. Prehistoric travellers may have attempted the crossing in coracle or canoe to avoid the acres of impenetrable forest and pathless bog that bordered the river. As he scrubbed off mud and sweat in the bath house, many a Roman centurion must have pondered the logistics of a direct route north from the camp overlooking the river at Cramond. The first known individual to cross regularly was the saintly Queen Margaret, wife of Malcolm III, as she journeyed between the royal palaces of Edinburgh and Dunfermline. Her son David I established a ferry around 1130 with free passage for travellers to and from the royal court. His grandson Malcolm IV first made the connection between queen and ferry when he granted the abbot and monks of Scone the right to cross the water free of charge at the *Fortum Reginae*. By the early thirteenth century Dunfermline Abbey, with its thousands of pilgrims to St Margaret's shrine, had acquired the rights to the Queensferry Passage. Individual 'oar holders' on both sides of the estuary held leases to operate the service, an arrangement that continued until the nineteenth century.

Ferry crossings were fraught, given the risk of storms, shipwrecks and seasickness. The south side lacked a fixed landing place until the early nineteenth century, with boats having to tie up against one of the rocky platforms along the shoreline depending on the state of the tide. Gangways for horses and cattle were only provided from the mid fifteenth century. Boatmen regularly took an additional cut for carrying cargo, refused to sail without a 'sweetener' or hijacked horses on board without their owner's consent. Official attempts were made to improve the service by, for example, ensuring that one boat was always available at each side of the Forth. Sunday sailings were banned except in emergencies. In 1749 a rudimentary timetable was introduced, governed by a bell rung from the public clock at North Queensferry where most boats tied up overnight. As soon as the first morning ferry was sighted boatmen were supposed to set off from the south side with a two-hourly service thereafter, regardless of whether passengers were waiting or not. Travellers in a hurry could catch an express service for a higher fare. Rushing to catch the northbound ferry, one unfortunate party hurtled down the steep Hawes Brae at such a gallop that carriage, horses and passengers overshot the pier and drowned.

In these days of high pressure, of living and working and eating and drinking at top speed, the saving of an hour or two for thousands of struggling men every day is a point of the greatest importance and every delay, however excusable and unavoidable, is fatal to enterprise.

Wilhelm Westhofen, 1890

Stormy crossing to North Queensferry
prior to the bridge being built

The ferry charges for 1749 show who was crossing and how much they paid at a time when a loaf cost a penny.

Each person crossing on the ferry in daytime	1s 6d
A boat express in daytime	£1 10s
A boat express 'under cloud of night'	£2
A yawl express by day	12s
A yawl express by night	18s
Coach or four-wheeled chaise	£3
Two-wheeled coach	£1 10s
Cart	12s
Covered cart or wagon	£1 4s
Each horse in a coach, cart or wagon	4s
Horse, ass, lowland ox or cow	4s
Highland or small ox or cow	3s
Sheep	6d
Swine	2s
Each horse-load	3s
'Cadgers with Fowls and other Vivers' (meat and fish sellers)	no charge for load

By the late eighteenth century the Forth was Britain's busiest ferry crossing, serving the Great North Road to Dundee and Aberdeen. Faster travel on improved roads and the growing demands of the postal service highlighted the unsatisfactory nature of the landings: on the south side, only Newhalls was operable at half tide or when winds blew from the north-west or south-east. In 1784 Queensferry town council noted the hazards to sailors when ferrying the mail. 'In winter they are compelled to scramble in the dark bearing a load of between 16 and 22 stones over a track of stones lying in disorder & covered with sea weed with sea water washing between them.'

Following a survey by eminent civil engineer John Rennie, improvements were finally made in the 1810s. North Queensferry gained a new pier at the Battery with separate waiting rooms for passengers and boatmen, as well as a house for the ferry superintendent and improvements to quayside cargo handling. Along the Queensferry shoreline four landings were created – Port Edgar, Newhalls, Port Neuk and later Long Craig. The Newhalls pier had a breakwater and paved roads on either side and six houses for boatmen nearby. Towers on each side of the estuary acted as both ferry signal stations and lighthouses for shipping in the Forth.

When James Watt built the prototype of his steam engine further upriver at Kinneil in 1769, few could have imagined the impact that his innovation would have on the Forth. By 1820 Henry Bell's steamers were heading as far as Stirling and the new route between Leith and Burntisland deprived the Queensferry crossing of two thirds of its coach passengers. The next year Queensferry struck back with its own steamer, the appropriately named *Queen Margaret*, custom-built to cope with piers designed for sailing craft. Ten new boats replaced the ageing sailing fleet. Nonetheless complaints about the service continued, one aristocratic passenger describing the 'heavy lumbering vessel' as 'the most miserable thing of the kind he ever crossed in'.

In 1863 the rights to the Queensferry Passage were sold to a new owner, the North British Railway Company. By now the skeleton of the Scottish Lowland rail network was complete. Edinburgh had been linked to Glasgow in 1842 and to London via both the east- and west-coast routes six years later. In 1850 the first London express steamed through Perth and Dundee on its way to Aberdeen. The five corporate giants of the northern rail network – the Scottish-based North British, the Caledonian and their southern counterparts, the Great Northern, the North-Eastern and the Midland – gobbled up the owners of smaller branch lines. They co-operated and increasingly competed to win the battle for a faster, more efficient and more profitable route to the North, the Caledonian being the only company not to back the Forth crossing.

The Forth and Tay estuaries barred the way. Goods and passengers for Dundee and Aberdeen had either to travel via Perth over the line of a hostile railway company or face two ferry crossings. The latter option was expensive, time-consuming, labour-intensive and, in the case of freight, potentially damaging. Railway engineer Thomas Bouch came up with an imaginative, if partial, solution in 1850 when he designed the world's first roll-on roll-off train ferry running from Granton to Burntisland. He contracted the famous Glasgow shipyard, Robert Napier and Sons, to build the vessel *Leviathan* with railway lines on deck. Ramps at each end of the five-mile crossing allowed wagons to run directly on and off the vessel while passengers travelled on a ferry alongside. The service ran successfully until 1890 when the opening of the Forth Bridge rendered it redundant. By then Thomas Bouch was dead, the victim of his

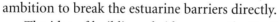

Bouch's Granton to Burntisland train ferry

Thomas Bouch

ambition to break the estuarine barriers directly.

The idea of building a bridge across the Forth was not new. John Grieve had put forward the first definite proposals in late 1805 with an attempt to raise funds for two tunnels west of Queensferry, 'one for comers and one for goers'. Edinburgh civil engineer James Anderson was the next to tackle the challenge with his suggestion of a chain bridge using the island of Inchgarvie, a third of a mile out from the Fife shore, as an anchor. Forth Bridge engineer Wilhelm Westhofen commented: 'The design would have given the structure a very light and slender appearance, so light indeed that on a dull day it would hardly have been visible, and probably after a heavy gale no longer to be seen either.'

Rail travel added a new dimension and sense of urgency as bridging the Forth would reduce journey times by an hour and a half. Three years before purchasing the Queensferry Passage, the North British had announced its chosen site for a bridge 6 miles to the west between Blackness and Charlestown, to a design by Thomas Bouch. With the river bottom proving too unstable to support a bridge and the North British in financial turmoil, the plans were abandoned, as was a later idea of a pontoon ferry that could directly carry trains. A train ferry service involving passengers having to change their mode of transport twice was eventually launched between North Queensferry and Port Edgar in 1878. The service did little to endear itself. The vessels' light draughts, dictated by the shallow water round the piers, meant that twenty minutes of misery was guaranteed in anything more than a stiff breeze.

The serious business of carrying the tracks across the estuary started in 1873 with the formation of the Forth Bridge Company, backed by the four railway companies with an interest in a direct route to the North. No one company alone could fund such a giant undertaking. Thomas Bouch designed a structure with two massive central towers rising from Inchgarvie and hung with a spider's web of steel chains. Journalists reassured nervous readers that a railway had long since been suspended above the mighty Niagara Falls and that the world's largest railway suspension bridge was currently under construction to join Brooklyn and Manhattan in New York. Bouch's design met general favour as 'light and imposing'. William Arrol was appointed contractor to the works.

A cotton spinner's son from outside Paisley, William Arrol (1839–1913) was the epitome of the Victorian self-made man. Sent to work in a cotton mill when only nine years old, he switched career in his teens to train as a blacksmith. Like many before him he then headed south for several years before returning to Glasgow to specialise in boiler-making. He set up William Arrol and Co. in 1868, opening the Dalmarnock Iron Works three years later and diversifying into bridge-building and public works. His growing reputation among railway company executives was partly founded on his inventiveness in designing tools to solve specific problems when building two railway bridges over the Clyde.

Arrol started by building offices and workshops on both sides of the Forth. His brickworks at Inverkeithing fired the materials for the first of the eight piers that

William Arrol

Artist's impression of Bouch's design for the Forth Bridge from above Queensferry

were to form the base of the Inchgarvie tower. Arrol was probably too busy to attend the laying of the foundation stone of Bouch's bridge on Thursday, 26 September 1876. The ceremony was kept very low-key because of the pressure to start work by the parliamentary deadline of 1 October. The party sailed out to Inchgarvie in open boats, the ferries being storm-bound. After Mrs Bouch did the honours with a gold trowel, the party retired to the Hawes Inn for lunch and speeches. The former provost of Dunfermline criticised the long delays which had jeopardised any proper rail services for his townspeople: a local MP claimed that Thomas Telford, father of the civil engineering profession, had designed a Forth Bridge in the 1820s. Mr Walker, the North British's general manager, hoped that Bouch 'would have good health and strength and be spared to see the consummation of what was really a great national undertaking'. Bouch laconically acknowledged the toast in a single sentence of thanks to his hosts.

The Tay Bridge, Bouch's first 'national undertaking' and the world's longest railway crossing, opened exactly twelve months later. The building of the Tay Bridge

had taken six years, ten million bricks, two million rivets and twenty lives. Nonetheless it quickly delivered profits to the railway companies and renewed prosperity to Dundee. Then, on the night of 28 December 1879, disaster struck. During an exceptional gale the bridge collapsed as a train with 75 passengers on board was crossing the high girders. There were no survivors. 'Appalling catastrophe', 'Dreadful tragedy': journalists and newspaper illustrators vied to convey the horror of the event. Two days later *The New York Times* summed up the story in bold capitals: 'The Tay Bridge Calamity; No One Left To Tell How The Accident Occurred. The Train Precipitated 88 Feet Before Reaching The Water – All The Central Spans Of The Bridge Gone – The Number Of Lives Lost Believed To Be Ninety – Only One Body Recovered.'

Once the Tay had ceased to wash up the flotsam and jetsam of the disaster – a leather slipper here, a blue-and-white ribbon and a black felt hat there – the public mood shifted from shock to recrimination and fingers increasingly pointed at Bouch as the scapegoat. The official inquiry, published six months later, concluded that his calculations had not made sufficient allowance for the strength of the winds that could rage down the estuary, as well as revealing a catalogue of contractors' short cuts, shoddy workmanship and lax supervision.

Bouch's status shifted from fêted knight to disgraced engineer in under a year. His son-in-law was a victim of the Tay Bridge disaster. Although Bouch suffered occasional bouts of ill health caused by a heart condition, the sheer stress of the aftermath of the Tay Bridge disaster, constant travel to London, the worry of maintaining his professional practice and the limelight of the inquiry focusing on a man who was shy and aloof by nature must have taken their toll. The loss of the Forth Bridge contract was the ultimate blow. Although he continued to work on the design of the Edinburgh suburban railway over the summer of 1880, his health finally broke and he died while taking the waters at Moffat spa two months later. Although officially his death was attributed to heart disease he was also described as suffering from 'a deep melancholia'.

Work on the Forth Bridge had immediately been halted after the Tay Bridge disaster. The railway companies were forced back to the drawing board as Parliament refused to sanction funds while Bouch remained in charge. The months passed with the project in limbo as the companies commissioned a committee of design engineers – William Barlow, Thomas Harrison and John Fowler – to explore every structural option and possible site from Bouch's original design to a tunnel. In 1881 John Fowler and his partner Benjamin Baker presented their radical proposals based on the principle of the cantilever to Fowler's fellow committee members. It was a winning design.

Yorkshireman John Fowler (1817–98) had dedicated his career to civil engineering, from waterworks and canals to railways and the London Underground's Metropolitan line. He rapidly reached the peak of his profession, being voted president of the Institution of Civil Engineers in 1865 at the unprecedented age of 48. Fowler first met Benjamin Baker (1840–1907) when working on the Victoria Bridge, the first railway crossing of the Thames into central London. Baker joined Fowler's firm around 1860, being made a partner 15 years later. Baker brought to the design of the Forth Bridge not only a deep understanding of materials, gained from his early training in a Welsh ironworks, but also an unusual talent for explaining complex, technical detail in a simple and engaging way. Although the two men put forward many proposals for long-span bridges in the course of their careers, including advising on the feasibility of a rail bridge across the English Channel, the Forth Bridge was the only long-span structure that they actually built.

The designers' brief was challenging. They had to span a mile-wide estuary whose bottom was solid rock or boulder clay. The depth of the main shipping channel around Inchgarvie was 200 feet, deeper than much of the North Sea. Tides and wind could be fast and fickle. The bridge had to make allowance for the largest naval or merchant ships passing underneath at high tide and for two steam trains passing at speed on the tracks. Given the anxieties of the travelling public since the Tay Bridge disaster, Fowler and Baker designed the structure to be strong enough to survive a hurricane at each end blowing in different directions. As American civil engineer Thomas Curtis Clark pointed out before visiting the site in 1887, it was crucial that 'the bridge had not only to be a very strong bridge, but that it looked very strong . . . When people looked at the bridge they would see that it was a bridge that no wind, no gale, no tornado could upset.' There were also commercial considera-

John Fowler

Benjamin Baker

tions to factor in: unlike the twentieth-century Space Programme, to which the gigantic undertaking of building a bridge three times the size of any other in the world can be compared, the cost-profit equation was only second to safety on the railway companies' agenda.

The cantilever was Baker and Fowler's chosen solution to this conundrum. A cantilever is a horizontal beam supported only at one end, leaving the other end projecting into space, like a theatre balcony or a diving board. Resting on its four piers each of the Forth Bridge's three towers was designed to hold a pair of 680-foot-long cantilever arms in balance, the diagonal struts above and below helping to take the strain. Each cantilever could be constructed independently, finally being linked together by two 350-foot-long girder bridges and the approach viaducts carrying the track.

In his design choice of cantilever and central girder Baker may have been inspired by a widely published, late-eighteenth-century sketch of an elegant wooden bridge in Tibet designed on the same principle. Others argue that Baker had only to look down his nose to admire the fine cantilever of his own moustache. In practice Fowler and Baker's choice was inspired. The position and scale of the central girder kept down the bridge's size, weight and hence material costs. It also minimised wind resistance at the central girder, the bridge's weakest point, and maximised it at the strongest point: the pillars.

Memorably, in 1887, Fowler and Baker created a human cantilever to demonstrate how the Forth Bridge worked. Swinging in the middle was no ordinary construction foreman but Kaichi Watanabe, one of the first Japanese students to train in Britain. He had joined the engineering practice on attaining a post-graduate qualification at the University of Glasgow. There his engineering professor was Henry Dyer, who had previously served as the first principal of the future University of Tokyo, being appointed in 1872 at the age of only 24.

'Experts and non-experts were certain the thing could never be realised. It seemed beyond the power of ordinary mortals.' So wrote 'An Intimate Friend of Mr Arrol' in a letter to *The Scotsman* in February 1913, long after the sceptics had been proved resoundingly wrong. Baker tested and retested his wind-stress calculations using a Heath Robinson-like contraption of a wooden rod, cardboard and string. Fowler asked the Ordnance Department to dispatch a team of military sappers to check his trigonometry in measuring the exact distance across the Forth. The results differed by one foot. Determined that the 1,710-foot-long spans should be accurate to an eighth of an inch, Baker rigged up posts, each exactly the same distance apart, along a straight stretch of railway line. He attached a length of steel piano wire to the posts, adjusting it to have a sag of 24 feet in the middle. The wire was then laid by steam launch across the width of the river, hauled to the surface, and adjusted for sag. The process was repeated several times while the officers and 300 crew of the Forth guardship HMS *Lord Warden* directed traffic in the estuary. Baker was satisfied that his measurements were correct to a quarter of an inch.

The first days of 1883 witnessed the start of the huge undertaking when William Arrol and the management team from his bridge-building joint venture, Tancred, Arrol and Co. moved on site just over three years after they had abandoned it. Their first action was to take over the hut and workshop that Arrol had built as part of Bouch's works, renaming it No. 1 Hut. In order to supervise the works directly,

Travellers approaching the ferry
at Hawes Pier

according to a friend, Arrol built himself 'a commodious villa' nearby. Site manager Joseph Phillips took up residence in Queensferry where he remained for the duration of the project. Staff recruited the first workmen, organised the repair of Bouch's workshops and chased boys off their recreation field, now needed for a cement-mixing plant and stores. Jetties soon pointed their long fingers into the estuary where over half a million cubic feet of finest Aberdeenshire granite for the piers and steel from as far away as Wales would be landed.

By the spring the landscape on both sides of the estuary was changing dramatically as the massive infrastructure took shape. Existing buildings were swept away, sites surveyed and foundations excavated. On the north side of the river, the rocky ground was levelled and the coastguard station rebuilt higher up the hill to make room for the railway piers. Joiners buried the centuries-old ferry landing at the Battery slip under timber staging where cranes were soon swinging loads of material from ship to shore and where boatload after boatload of Briggers would soon disembark at the end of their shift. Preparations were even more extensive on the south side where much of the prefabrication was to take place. A 27-acre field above the Hawes Inn had been commandeered and the cliff-like slope blasted to create terraces for the foundations of offices and workshops. Men laid a siding off the main railway line where the half-finished girders were to be craned off the wagons and manhandled into the workshops. The tracks between Queensferry and Dalmeny were doubled to cope with the anticipated volume of traffic. As operations intensified, land on both sides of the railway was gobbled up, two crossings being created by girders rescued from the Tay Bridge.

A 100-foot-long drawing loft was erected on whose blackened floor draughtsmen would mark out full-scale drawings for the engineers' approval. From these drawings carpenters would create and brand wooden templates of every component before it was shaped in the furnaces, drilled or planed. The site of an enormous shed was staked out, large enough to allow for the completed girders to be laid out in exactly the position that they would occupy on the bridge. Foundations were dug for the building to house the gas-fired furnaces that heated the steel to be manipulated into girder tubing. Ground was levelled for the drill roads where tubes would be assembled. Machinery arrived on site daily, much of it designed by Arrol himself specifically for tasks like drilling and riveting. Telephone communications were set up between offices, workshops, stores and the three main construction sites – Queensferry, Inchgarvie and North Queensferry – by running a cable under the Forth. By the late 1880s girders and tubes, like giant jigsaw pieces waiting for the final picture to emerge, spread out over 50 acres of land. Essentially, the Forth Bridge was built twice, once on land and once at sea.

On 14 November 1883 electricity arrived in Queensferry when 16 Swan incandescent lamps were turned on in the work sheds. Gas lighting had been ruled out because of the 'ruinous rates' charged by Queensferry's gas company. Given that Swan had only patented his light bulb three years before, the *Dunfermline Journal* was right to be impressed: 'So far, it appears to be very successful, as every article in the sheds, from one to the other, could be clearly discerned.'

Six months later Charles Henry Farley Cox cycled the 393 miles from Norwich to Queensferry on his penny farthing to take on a new job as foreman responsible for

Engineers with surveying instruments; Westhofen eighth from left

The drawing loft

electrical installations on the bridge; he took three weeks to make the journey. Eighteen months later he was back on his penny farthing heading for home.

Over the first year, a 2,100-foot-long jetty gradually snaked its way from the shore to the site of the Queensferry pier, its planking rising 8 feet above high water. A pile-driving barge working from the seaward end was brought in to speed up operations. At the landward end, a sloping roadway rested on timber trestles, crossing the Edinburgh road on an iron-girder bridge. From water and fuel oil to machinery and components, each coated in boiling oil and stencilled with its final location, everything needed for offshore working travelled down the roadway in wagons which were then hauled back up by cables. At the foot of the pier steam cranes manhandled equipment onto a fleet of four steam launches and eight steam barges, each capable of towing a string of vessels behind it. Only the basic building materials such as Arbroath freestone for the rubble infill of the piers arrived directly by sea. Rafts of timber logs were floated from the port of Grangemouth to be processed in the sawmill downriver from the Queensferry site.

Preparatory work on Inchgarvie, the island which the Forth Bridge Company had bought from the Dundas family, lasted a year and a half. In summer 1883 wind gauges were mounted on the late-fifteenth-century castle, built to protect the Forth from pirates and used over the centuries as an isolation hospital, state prison and fortification against attack during the English civil war and the American War of Independence. Nine months later, the castle and its battlements were roofed in to provide essential stores and offices. Where possible the rock was levelled as every inch of space was vital on a site which would become the base for the 900 men working on the Inchgarvie cantilevers. A 10,000-square-yard cloak of iron staging was thrown over the island, divers pinning it to the seabed. It was reinforced to ensure that it could take the load of tons of components and machinery. Until the timber stages and jetties were built only small boats could moor, resulting in a significant loss of time. Later, during the construction of the north cantilever, strong currents and a difference in tidal levels of up to 22 feet made it next to impossible to hold the steam barges in place for long enough to attach the lifting tackle to unload material.

Work started in earnest in summer 1883 with the laying of the foundations for the piers of the railway viaducts and those cantilever piers which rested on land, including the northern legs of the Inchgarvie pier which were to sit on the rock between the tidelines. For the base of each leg a circular groove 3 feet deep and 60 feet in circumference was hewn out of the rock into which an iron belt was hammered. Where the rock sloped a shield closed the gap. A temporary caisson, like a huge enclosed cylinder, was built on top to weigh down the belt until divers could pack the joins with bags of concrete. These giant cylinders were known as half-tide caissons because over 250,000 gallons of sea water had to be pumped out of them between every tide. Once the foundations were filled with rubble masonry, building the piers themselves could began. While many of the viaduct piers on the south side rested on the hillside or on rocks exposed at low tide, half-tide caissons had to be brought in where solid rock was encountered, as on the Fife shore where the rock had first to be levelled by blasting. Inevitably progress was slow as the Briggers were constantly buffeted by wind and waves, work having to be halted entirely during spring tides and rough weather.

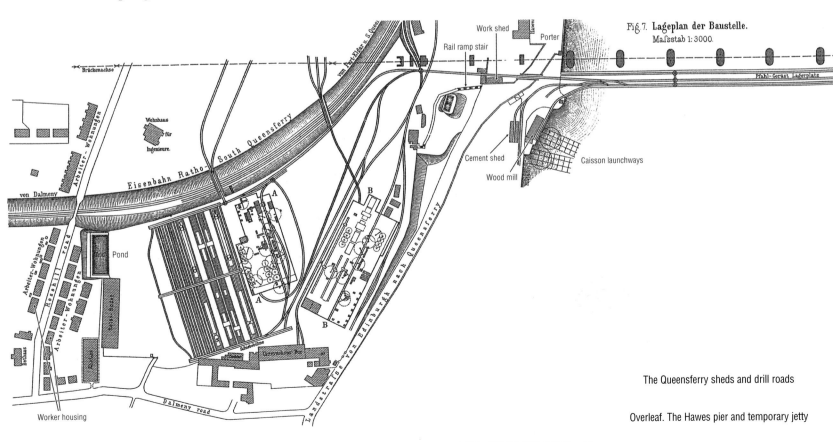

The Queensferry sheds and drill roads

Overleaf. The Hawes pier and temporary jetty

The pneumatic caissons were the monsters which most of all captured the public's imagination. They were said to be the largest ever built in Britain. These six caissons were used to create foundations on the seabed for the Queensferry cantilever and the southern Inchgarvie piers, coffer dams being utilised to create the other foundations. The position of the Inchgarvie piers had first been confirmed by taking over 3,000 soundings from a raft, 70 feet in diameter, floated out from Queensferry. Everything from air compressors to mounds of broken stone was prepared in readiness for the arrival of the caissons.

Each caisson consisted of a wrought-iron cylinder 70 feet in diameter and from 45 to 71 feet high after sinking, which tapered at the top. Inside was a second cylinder, the space between the two walls being divided into compartments which could be filled with concrete to weight the caisson and control its descent. A steel rim was attached to the base of the 7-foot high cone-shaped working chamber. Pumps maintained the internal air pressure at a sufficient level to stop water entering from below. This space was kept free of water by pumping in air. It acted as a diving bell allowing the workmen to clear rocks, gravel and mud from below the rim of the caisson so that it could sink into the river bed.

Air shafts, each only 3.5 feet in diameter, rose vertically through the cylinder providing access to the outside world. One or, in some caissons, two shafts were used for the removal of rock and debris while pipes, through which a small amount of compressed air was allowed to escape, jetted out sand and silt in a stream of water. The air forced through the chambers ended in a mass of bubbles making the water surrounding the caisson appear to be boiling. Hydraulically operated airlocks at the top of the shafts allowed materials to pass in and out without releasing the air

Briggers working at the drill roads

THE FORTH BRIDGE

Sinking of Inchgarvie caissons

pressure. The men clambered down another shaft to their workspace by ladder, plans to install a passenger lift having been abandoned. They operated by hand the valve of the airlocks to allow air in and out without the pressure dropping too abruptly.

Like great ships, the caissons were launched on cradles from the slipway built in a sheltered bay east of the bridge. They had been manufactured by Arrol Brothers of Glasgow, unrelated to William, dismantled and transported to Queensferry for re-assembly. On 26 May 1884 the Countess of Aberdeen wished the first caisson 'God speed' as, towed by tugs, it made the short crossing to the south-west pier of the Queensferry cantilever. The crowd gasped as the caisson listed and then righted itself with the sound of what one onlooker described as an 'unexploded torpedo' when air trapped in the lower chamber was released. The Briggers enjoyed an unexpected half-day's holiday. At the launch of the third caisson a real torpedo made its presence known, exploding during a training drill by sailors from HMS *Lord Warden*.

Once in place the caissons were prepared for action. Temporary caissons were built on top to protect the working areas from flood tides. Pumps, cranes, concrete-mixers, engines and cement ballast to sink the caissons into their foundations were unloaded onto work platforms. Great care had to be taken to avoid sinking the caissons too quickly, thus filling the working area with mud and trapping the men, such accidents having happened elsewhere. On the way down the exact position of each caisson was checked and checked again. Even if a fraction out of line, the caisson was guided back on course, a very delicate operation given its weight. When the boulder clay at the bottom proved difficult to work, Arrol, who always rose to a challenge, designed special hydraulic spades to remove it. Braced against the roof of the working area, the spades could dig down into the clay and also under the rim as it was lowered into the clay, cutting out square slabs which the men then loaded into buckets for disposal. Once the caisson was lowered, however, the clay provided an effective seal allowing the air pressure to be reduced and making working conditions easier despite the great depth.

On New Year's night 1885 the north-west South Queensferry caisson, which was being fitted out prior to sinking, grounded on an exceptionally low tide and stuck firmly in the mud. Three months later, during an attempt to refloat it, the caisson tipped over and filled with water from the rising tide. The men pumped the water out too quickly resulting in the water outside bursting through and two deaths. It took a further seven months of patching with timber and clay before the caisson, clad like a gigantic cask, was in a fit state to be refloated prior to being sunk onto the seabed.

Inchgarvie's sloping seabed made life particularly difficult. Within the diameter of the caisson itself there was a 20-foot slope. When one side of the caisson came to rest on the rocks, the other had to be supported with thousands of sandbags filled with concrete to keep it in balance. They were thrown down from a raft over the top and adjusted by divers. The caisson also had to be propped with wooden wedges to stop it drifting on its descent. The workers had to penetrate hard rock, rather than the layers of silt, mud, pebbles, stones and boulder clay encountered on the southern shore. At first teams of three men used hand drills to create holes for explosive charges; later Arrol came to the rescue and designed pneumatic rock drills which speeded up progress. Before the charges were fired, the men withdrew and a wooden trapdoor closed the bottom of the air shaft to prevent the fumes from rising. Air compressors then cleared the chamber of fumes. Firing often took place at the end of a shift to keep lost time to a minimum. As the rock was blasted and chipped away, the debris was passed up the air shaft in buckets. After many weeks of hard work the caissons finally rested on a platform of solid rock. With the caissons securely anchored the workmen gradually withdrew, filling the chamber and shafts with concrete as they went.

Resting on their caissons, the 12 circular granite piers broke through the waters of the estuary. At either end the 150-foot-high masonry viaducts rose skywards in step with their girders. Hydraulic jacks lifted each end of the girder, a maximum of a foot at a time, always keeping the girders 6 feet above the masonry being laid below it. Building the viaduct and jacking it up in step with the rising piers was an ingenious solution to the problem of how to place the viaduct on top. With their scaffolding platform and travelling rails the girders themselves acted as mobile

Opposite top.
Filling a caisson with ballast prior to sinking

Opposite below.
Pumping out the tilted caisson

workshops. At first lifting a girder took a day or more but with experience the time was slashed to three and a half hours. Soon the pattern was established – lift the girder 3.5 feet, build two courses of masonry, lift the girder 3.5 feet, build two courses of masonry – this routine continued for 15 months until July 1887 when the men hoisted flags along the girders to celebrate the viaduct piers reaching their full height of 130 feet.

By now erecting the central towers and building out the six pairs of cantilevers was well underway. Their sheer size earned the structure the title of 'The Giant Bridge'. First of all the Briggers bolted the bedplates, several layers of metal thick, onto each pier, the 48 bolts being embedded 26 feet into the masonry. Skewbacks, like huge upturned hands, were bolted onto the bedplates to carry the cantilever towers. Only one of the four skewbacks of each cantilever was rigidly fastened, the others having slotted holes to allow the superstructure to move as it expanded and contracted with changes in temperature. In practice the bridge only briefly 'moved' as it proved too difficult to lubricate the sliding bedplates.

Then the miles of tubing started to spread their fingers upwards and outwards. Their plates had first been bent into shape and assembled on the drill roads beside the works. Special machines drilled the rivet holes. Once checked for accuracy, each plate was dismantled and shipped out to where it was needed. Verticals, diagonals, struts, braces and ties started to create the patterns of lattice and diamond. Cranes hoisted the pieces from barges and crawled over the superstructure, held in place

only by their 50-ton weight. Their jibs swung constantly backwards and forwards as they fed materials to satisfy the appetite of a monstrous chick. Cages of riveters hung from work platforms as they inched their way up the towers. Miles of steel rope held the structure together temporarily until the rivet gangs had completed their work.

Only towards the end of the building of the cantilevers were many of the public able to understand where, within this apparent chaos of metalwork, the railway would actually run. By the end of September 1889 the ends of the girder between the Queensferry and Inchgarvie cantilevers were so close that they could be bridged by a gangway. There was serious competition among the Briggers to be the first to cross. The winner was the workman who had the idea of positioning cranes on top of each half of the girder and placing a ladder between their jibs. A Mr Campbell, in charge of the northern girder, was the first person to cross officially once the gangway was in place. That night the health of the engineers and contractors was toasted in the Hawes Inn. The final step was to bridge the short gaps between the cantilever ends with the girders that carried the rail track. The ultimate test must have given the engineers sleepless nights. Would the cantilever arms link together? They did. The north span was only one inch and the south span six inches shorter than on the original plans thanks to the meticulous measurements of the surveyors. Their skill also resulted in an error of only two inches in the six cantilever arms which stretched out for nearly 700 feet. The fact that this error was always to the east led to speculation that it was due to prevailing winds and temperature rather than inaccurate calculations.

On 21 January 1890 a strong gale blew as, side by side, two trains entered the bridge from the south. This was the first time that the bridge had felt the weight of what it was designed to carry. These were no ordinary trains, each consisting of two locomotives, 50 wagons weighing over 13 tons and another engine at the rear. The engineers had already anxiously studied their rules and breathed a sigh of relief. The amount of movement of columns and cantilevers could be measured in inches, well within the Board of Trade's requirements. Six weeks later, on 4 March, the Prince of Wales stepped from the saloon of the royal train and secured the last of the 6.5

Fabricating lower part of cantilever, at the drill roads

Opposite. Starting to build a cantilever

million rivets. The train then steamed on to Queensferry and an official reception, while a pigeon, released during the ceremony, was blown downriver towards the capital.

Now the bridge had to be tied into the rail network. On the south side this was a relatively easy task thanks to the local topography. The land behind the cliff that dropped steeply down to the shore was at the same level as the viaduct onto the bridge. The very different terrain on the north side meant that the new line had to be taken through cuttings and tunnels and on a viaduct for nearly 2 miles to reach Inverkeithing. The greatest challenge, however, on this stretch, turned out not to be blasting through rock or excavating tunnels but stabilising a bog to take the weight of an embankment. At one point the ground shifted sufficiently to move the nearby public road 60 feet.

The rail network spread its tentacles from the bridge in all directions. New stations were built at each end, the one on the south side being honoured with the name Forth Bridge Station. It was later renamed Dalmeny. A line linked Dalmeny Junction to the North British's main Glasgow–Edinburgh line both at Edinburgh's rapidly growing western suburb of Corstorphine and at Winchburgh, thus creating a route from Glasgow to the North that avoided the capital entirely. Bouch's Granton ferry was finally rendered obsolete with a new line from Inverkeithing to Burntisland and onwards by the existing route to Dundee and the North. Travellers from Edinburgh to Perth no longer had to detour via Stirling thanks to a scenic route through Glenfarg. There was talk of London trains steaming into Aberdeen in just over twelve hours.

The bridge finally opened for limited passenger business on 5 March 1890, although the main route north did not operate until the final links of the network

Opposite. Erecting tubes on the Fife cantilever showing the riveting cages

Below. North rail approaches viaduct at Jamestown, Inverkeithing

were in place by early June. A March gale blew and the first train rocked as it crossed but the bridge itself barely vibrated, the final proof of Fowler and Baker's vision. All that remained was to dismantle everything that was not a permanent feature of the bridge, from the staging round the piers and landings to cranes, cement mixers and miles of wire rope. In July workmen erected brass memorial tablets on each side of the track above the arch near the south cantilever. One recorded the date of the start and opening of the bridge, the other the names of the railway directors, engineers and contractors. There was no memorial, however, to the Briggers who made the eighth wonder of the world possible.

Opposite. Joining the cantilevers

Below. Phillips compared the bridge to other world icons

The Victorians had an insatiable appetite for statistics.

	Feet	*Metres*
Overall length	8298	2529
South approach	1980	603
North approach	970	296
Cantilever length	680	207
Main spans	1710	521
Maximum bridge height	361	110
Rail level above high water	156	48
Navigation headway	150	46
Steel used	55,000 tons	55,885 tons
Masonry	140,000 cubic yards	107,038 cubic metres
Total cost		£3.2m

CHAPTER 2

BRIGGERS, BOSSES AND THE BLACK SQUAD

An operation as vast and complex as the bridge demanded not only obedience and endurance but also brains and brawn. The thousands of workers directly involved in its construction included ambulance drivers, bricklayers, cooks, concrete mixers and crane drivers; design engineers, draughtsmen, divers and demolition gangs; electricians, fitters, foremen and gangers; holders-on, inspectors, insurance clerks and joiners; key workers, leading hands, metal testers and navvies; overseers, painters, pattern cutters and platers; quarriers, riveters, surveyors, storekeepers and stonebreakers; template cutters and tunnellers, unloaders and valuers; wages clerks, watchmen, watermen and yard workers.

The bridge also impacted indirectly on many other lives. Lawyers representing the railway companies composed the fine print of contracts and resolved disputes. Journalists penned superlatives from their desks in Dunfermline or New York. The armed forces played their part, from the sappers who assisted Baker with his measurements to the sailors on HMS *Lord Warden* patrolling the Forth. Orders for 42,000 tons of steel provided work for hundreds of men in Lanarkshire and South Wales while a firm in Devon was one of four suppliers of paint. Innkeepers on both sides of the Forth were kept busy pulling pints and pouring chasers. Even prostitutes found a new outlet for their services.

Although he had formed a joint venture with Sir Thomas Tancred, aristocrat, consulting engineer and one-time New Zealand sheep farmer, the bridge was William Arrol's baby. Like a mother he devoted his energies to it day and night, while also overseeing the second Tay Bridge and contracts in London. A close friend commented on Arrol's death in 1913: 'Only an iron constitution and an indomitable will could have carried the great engineer through these anxious, bustling years.'

His weekly routine was formidable. Rising at 4 a.m. on Monday mornings, he was at his desk at Dalmarnock Iron Works less than an hour later, looking over plans and checking the detail of work in progress. After a hurried breakfast in Miss Cranston's tearoom he caught the Edinburgh train to Corstorphine where a special engine waited under steam to take him to Queensferry. A day of meetings often ran well into the night; no detail was too small to escape Arrol's unerring eye and no

The able men who have built the bridge have not forgotten the willing obedience, the patient endurance, the high courage of those who have laboured on the construction of it.

The Scotsman 4 March 1890

Constructing a cantilever bedplate

mechanical problem too challenging to tax his ingenuity. Until its opening in June 1887 Arrol spent Tuesdays overseeing progress on the Tay Bridge, returning to Glasgow late in the evening. Wednesdays and Thursdays were spent at his desk before catching the sleeper for a round of meetings in London with the design and civil engineers on Fridays. Saturday saw him back in Glasgow, often working well into the evening. For Arrol even Sunday was far from a day of rest. After church he regularly walked the 14 miles to Paisley and back to visit an ailing relative who had been kind to him in his youth. Despite this hectic schedule Arrol found time to build himself an Italianate palazzo called Seafield on the outskirts of Ayr. With its wood-panelled library, billiard room and conservatory, housing palms and tropical plants, it was very different to the room-and-kitchen in Glasgow's east end where he lived as a young man.

Thrifty, thorough, inventive, practical, honest, earnest, alert, self-reliant – few people had a bad word to say of this blunt-spoken man of the people who trained his assistants largely by example. He was admired for his common sense and his good business head. Thinking nothing of working a fifteen-hour day himself, he was puzzled by his workers' view that an eight-hour day was more than long enough. In broad Scots he would chide them by recalling the struggles of his youth.

Working under Baker, Fowler and Arrol were the experts like Allen Stewart, who had worked with Bouch on the Tay Bridge. Patrick Meik, the resident engineer on the foundations, was known for his 'unfailing tact and courtesy'. Meik came to the notice of Fowler and Baker while constructing harbours like Bo'ness and Burntisland on the Forth, under the supervision of his Edinburgh-born father Thomas, who founded an engineering consultancy in 1868. Meik's firm still trades today as the international

Forth Bridge directors with Fowler and Arrol at the drill roads

Halcrow Group. William Gray, in charge of excavation and masonry, shared Arrol's appetite for work: 'Day or night, early or late, no one would ever call upon him and find him unwilling to do what was wanted.' Joseph Phillips had built his reputation as a specialist in glass-and-iron buildings, including the Crystal Palace and the roof of Birmingham New Street, which had the widest span of any railway station in the world. Working on the Rochester Bridge, the first to use pneumatic caissons, gave him the experience to work closely with specialist sub-contractor Monsieur Coiseau of Paris on the Forth Bridge caissons, as well as assisting Arrol with the super-structure. The bridge also provided a launch pad for the careers of several young civil engineers including Sir Ernest Moir, who went on to patent the Moir pill box used by gunnery troops in the trenches at the end of the First World War.

Managing the operation was a gargantuan task. Materials had to reach the place they were needed at the right time, requiring, according to Westhofen, 'the exercise of energy and skill as well as tact and patience, for the work was equally pressing on all points, and no one liked to be left behind in that great race for supremacy'. The programme slipped badly in the early stages, as until the jetties and staging were built

only the smallest boats could operate at low tide. Plans had to be changed in the light of missed deadlines. Managers were constantly on the lookout for ways to cut time and costs. By his construction methods, former Greenock shipyard worker Adam Hunter, in charge of the Dalmeny workshops, reduced the estimated time of the erection of the bridge by two years. He was rewarded in 1890 by being appointed resident engineer.

It is thanks to the engineers that we know so much about the construction of the bridge. German-born Wilhelm Westhofen (1842–1925), in charge of constructing the middle cantilever, wrote an account for *Engineering Magazine* in 1890. Profusely illustrated with technical drawings, its 70 pages of densely packed type provide tantalising glimpses of the working lives of the men among the reams of statistics and data on construction methods. Photographer Philip Phillips, son of Joseph, the bridge's resident engineer, had a bird's-eye view from his Queensferry home at Catherine Bank, taking pictures at weekly or fortnightly intervals during 1886 and 1887. Phillips had a sense of humour. He doctored one photograph to show how the bridge would look when finished and on another, sketched in a train steaming along the track. Cashing in on the public's fascination with the increasingly iconic structure, he published several accounts illustrated with his own photographs, including *Sketches of the Forth Bridge, or, The giant's anatomy, from various points of view.*

As an assistant engineer and the bridge's official photographer, Evelyn Carey (1858–1932) unusually combined a creative eye with a deep technical understanding of his subject matter. No other photographer was allowed access to the unfinished bridge. Commercial specialists like George Washington Wilson and James Valentine, from whom Carey sought professional advice, had to confine themselves to taking photographs from the shore. Given the cumbersome equipment of the day, Carey faced formidable challenges in taking shots from inside the cantilevers. Even more remarkable were his photographs taken inside the air chambers of the caissons, which required up to 20 minutes' exposure. Carey also had to cope with the uncertain light of the arc lamps and atmospheric changes. When the air pressure increased even slightly, the atmosphere cleared and the air rushed out under the caisson edge with a noise like distant thunder. A huge wave of cold water came rushing back, obscuring the air chamber for a time with a dense white fog.

Carey was a pioneer of industrial photography. His images provided the public with a realistic and accurate portrayal of the construction of the bridge, unlike the engravings widely used in magazines such as the *Illustrated London News*. Their artists showed little understanding of the bridge's structure and set it within an idealised landscape. In sympathy with Benjamin Baker's vision of a bridge that was beautiful as well as structurally sound, Carey's distant views from vantage points like Blackness Castle sought to show that great human endeavours were in harmony with the natural world.

With the magnification of modern digital technology, Carey's huge glass plates assume a new significance. For the first time the tiny ants clambering over the structure on Carey's prints can be revealed as individual Briggers, working, at rest or posing for the camera. The images turn the construction of the bridge from a technical achievement to a human story. Who were these men, where did they come from, what were their lives like, what did they think of as they hammered in rivet after rivet or manhandled the plates on the drill roads? Birth and death registers, health and police records, newspaper accounts and family heirlooms provide glimpses that draw a few of the men out from the shadows of their bridge. Many questions remain unanswered: many lives are lost for ever.

Bricklayers and joiners were among the first to be recruited to make good the buildings associated with Bouch's original works; by mid April 1883, 200 men were working on the south side and 100 on the north side of the river. Eighteen months

Revealed for the first time

Using hydralic riveter on the drill roads

later the workforce had risen to 1,200. It peaked in the summers of 1887 and 1888 with a total workforce of over 4,500. Numbers fluctuated depending on the stage of the operation and the season, with many hundreds getting their 'red cross', a sign that they were being laid off, as winter approached.

Most Briggers were labourers, ranging in age from early teens to late sixties. Some would have experience of other major construction projects as the 1880s was a time of laying down new infrastructure – bridges, railways, ports and docks. Some had previously worked on Bouch's ill-fated Tay Bridge. Men regularly travelled with the work wherever it happened to be and Arrol, as a good employer, must have attracted a loyal following. There were a few ex-jailbirds 'too clean shaven or too close-cropped as to the hair' and con-artists who bragged about their last big job. Some of the riveting teams were familiar with the scale and hazards of the bridge from working high up on the hulls of liners and warships on the Clyde. The bridge provided them with a welcome source of employment during the 1880s recession in shipbuilding. Agriculture too was depressed, possibly accounting for the number of men originating from the rural north-east of Scotland and the eastern counties of England. More locally the bridge drew on labourers from Leith docks and casuals from Edinburgh's Old Town.

According to Westhofen, by nationality the men were almost equally divided among the Scots, the English and the Irish, with the last being largely unskilled albeit 'hard working and conscientious'. Men from poor rural areas like Donegal caught the regular steamer service from Derry to Glasgow. Some stayed for the duration while others decamped after a week or two. 'Many of them – hundreds of them – were mere birds of passage, who arrived on the tramp, worked for a week or two, and passed on again to other parts, bringing a pair of hands with them and taking them away again, and having in the meantime made extremely little use of them except for the purpose of lifting the Saturday pay packet and wiping their mouths at the pothouse.' The Irish contingent also included men from families who had already settled locally, working in Leith docks or in the West Lothian shale mines.

Admissions of men living in Queensferry to Edinburgh Royal Infirmary as a result of an accident during the period between 1883 and 1887 gives an indication of the geographical spread of the workforce. Most were Briggers although some 'labourers' may have been working in the shale mines and quarries or on local farms. Surprisingly few came from Fife or West Lothian.

Occupation	Place of Origin
	Scotland
Engine driver	Tayport
Labourer	Galashiels
Driller	Dumbarton
Fireman	Perth
Riveter	Dunfermline
Labourer	Aberdeen
Labourer	Glasgow
Fitter	Stirling
Labourer	Glasgow
Labourer	Fochabers, Moray
Labourer	Keith, Banffshire
Iron foundryman	Airdrie
Labourer	Banff
Shipwright	Govan, Glasgow
Machineman	Glasgow
Labourer	Inverness
Labourer	Fraserburgh
Labourer	Perth
Scissor grinder	Inverness
Plater	Paisley

Occupation	Place of Origin
Engineer	Bridge of Weir, Renfrewshire
Engineer	Falkirk
Riveter	Dundee
Labourer	Haddington
Driller	Glasgow
Tinsmith	Aberdeen
Labourer	Glasgow
Labourer	Old Monkland, Lanarkshire
Labourer	Dundee
Labourer	Partick, Glasgow
Labourer	Dundee
Riveter	Elgin
Labourer	Grangemouth
Fitter	Glasgow
Engine fitter	Montrose
Boiler maker	Pollokshaws, Glasgow
Painter	Glasgow
	England and Wales
Signalman	Lancaster
Labourer	Stafford
Labourer	Stockton on Tees
Labourer	Lincolnshire
Labourer	Sheffield
Plater	Newcastle upon Tyne
Labourer	Spalding, Lincolnshire
Labourer	England
Fireman	Holyhead, Wales
Labourer	Wales
Labourer	Boston, Lincolnshire
Labourer	Chester
	Ireland
Riveter	Ireland
Painter	Dublin
Labourer	Belfast
Labourer	Londonderry
Boiler maker	Ireland
	Overseas
Engineer	Italy
Labourer	Belgium

The caisson workers were a race apart. Coming from France, Belgium, Austria, Germany and especially northern Italy, like a travelling circus they moved around Europe from job to job wherever their skills were required. Under the management of Monsieur Coiseau they arrived in the Forth from Antwerp, where they had spent several years laying the foundations of the harbour and docks. Some had also worked on the Suez Canal and on the Danube river works around Vienna. The first to arrive were around 20 Italians in May 1884 to work on the foundations for the Queensferry cantilever. According to *The Scotsman*, 'It is understood they can stand this high pressure much better than British workmen.' By October 1885 most had left for other projects, a few being retained to repair the damaged caisson. Foreign workers were not only among the first to be employed but also among the last as the laying of the asphalt footpaths along the side of the railway track was sub-contracted to the Seyssel Asphalt Company who mined bituminous asphalt in the French Jura mountains.

Housing the men presented an immense logistical challenge. Initially it was envisaged that most would be accommodated locally. In the summer of 1884 all but 200 of the 1,200 workers were resident in the Queensferries. Some staff and foremen moved into 16 purpose-built houses in Rosshill Terrace. Perched high above the Queensferry shore and nicknamed 'the Brickies', 60 tenement flats in Catherine Terrace housed leading hands and gangers. On site visits Fowler stayed in the semi-detached *Forthview*, its twin bay windows overlooking the drill roads, while Bridge House was reserved for senior engineers. The unprecedented demand for housing pushed up rents by 100 per cent in Queensferry and by over a fifth in places as far away as Kirkliston and Dunfermline. Many Briggers lived in the 'huts', wooden dormitory camps on both sides of the river whose facilities included canteens, reading rooms and stores selling boots, clothing, food and groceries. Local people took in lodgers, packing them in wherever there was floor space. After a serious outbreak of scarlet fever in North Queensferry in the autumn of 1887, the sanitary inspector recommended that a cleaner was employed and that overcrowding in both lodging houses and the huts was addressed.

The foreign caisson crew was segregated from the rest of the workforce to be close to the job, especially given that caisson disease, now known among divers as 'the bends', usually came on within half an hour of exiting the airlock. Commuting from Edinburgh would also have presented difficulties as few spoke English. At the start of operations the contractors built a 'village' for ninety men on Inchgarvie including 'a cottage and a kitchen'. The rotting hulk of the sailing ship *Hougoumont*, previously used as a cement store, was proposed as an offshore dormitory for the Belgian divers but they refused to move in.

The *Hougoumont* had a notorious career. Built in Burma in 1852, she was originally part of the fleet of ships owned by Duncan Dunbar, who provided nearly a third of the convict transport to Western Australia. In 1867 she became the last ship to carry convicts to Australia including 62 Fenian political prisoners. By the time the Forth Bridge contractors purchased her she was little more than a hulk. Her new cargo was up to 1,200 tons of Portland cement, which had to be stored for a set time after ships brought it from the manufacturer in Kent.

After it became obvious that local housing arrangements were totally inadequate the North British laid on special trains from Edinburgh, Dunfermline and Inverkeithing. Fares were low – 2s for a weekly season ticket from Edinburgh or 2d

Taking soundings from surveying raft for Inchgarvie caissons, the *Hougoumont* in the background

for a single – with the men from Fife paying about half the price. The contractors deducted the fares from the men's wages. There were two morning and two evening trains running one behind the other to take the Briggers on and off shift. The service was later extended to Leith where men had to rise at 4 a.m. for the day shift, returning home at 7 p.m. When the weather turned too wet or stormy to continue working, the trains were summoned by telegraph to take the workers home rather than risk them using the local pubs as a waiting room. An additional summer option was to commute by steamboat which Westhofen considered to provide 'an enjoyable and healthy trip for the tired workmen' who 'preferred this to the other alternative of living in the overcrowded rookeries of the neighbourhood'.

A flotilla of small boats ferried the Briggers to the bridge every morning and back at night, as the men did not necessarily work on the side of the Forth nearest home. Timing was critical to match the ferrying operation with the Edinburgh and Dunfermline trains. A paddle steamer was initially hired until a dedicated vessel could be built with the capacity to carry 450 Briggers. Decked-over steam barges, the steam tugs *Dolphin* and *Ruby* (until she ran aground in August 1885), the steam launch *Dart* and a fleet of rowing boats supplemented the paddle steamer, which between commuter trips shuttled across the Forth every hour carrying works personnel and visitors. Until security was tightened it also became a popular ride for tramps and others wanting to hitch a free lift across the Forth. Safe passage depended not only on the weather: 'There was always a number of unruly and reckless men whose conduct brought mishap and injury on others as often as on themselves.'

As well as the works canteens there were shelters and dining rooms on the three erection sites, on top of the central towers, at the level of the viaduct and right out near the ends of the cantilevers. The men could retreat to the shelters during sudden squalls or showers as well as meal breaks with a man being employed to heat up food on a stove. Drawn from the Dunfermline water supply, drinking water was shipped to Inchgarvie in iron boxes enclosed in wood. As there were no toilets the Briggers must simply have relieved themselves over the side.

Caisson workers occasionally escaped the monotony of the canteen diet by catching salmon, dogfish, octopus, crabs and lobsters which were attracted by the lighted chamber. One salmon weighed ten pounds. In a frantic attempt to escape, one fish leapt nearly the height of the chamber before jumping into an empty skip. Its final fate was a pot of boiling water.

Left. At the end of the shift the Hawes Inn beckoned

Overleaf. A workmen's hut on top of a cantilever

Saturday was pay day. The contractors were known to award an advance to certain workmen on the recommendation of their heads of department, leading to anticipation and often then disappointment. In June 1887 the *Dunfermline Journal* reported: 'Only in a few cases on Saturday was an advance conceded, and great dissatisfaction was expressed by a large body of the workmen.' Overall pay and conditions were good. Wages were in general higher than for the average labouring jobs in construction and agriculture, especially during a time of economic depression where pay in many sectors had fallen considerably. A labourer could earn 4.5d to 5d an hour, an engineer 6.5d and a carpenter or plater 7d, but, at least initially, there was no extra 'danger money', a subject of dispute between men and management. Many trades were paid piece rates, the Inchgarvie caisson workers' earnings, for example, being measured by the number of cubic yards of rock removed. Riveters could double or treble the normal labourer's wage, taking home an average of £2 a week or £3 if the team was working flat out. They often divvied out the take in the local pubs where members of the gang also held kangaroo courts if someone was not pulling their weight.

Training was largely on the job, although management was aware of the need to allow time to build the men's confidence. One advantage of a cantilever design was that it rose upwards and outwards from its base, permitting the men to gain experience before moving to the most dangerous work environments. According to Westhofen, the Briggers 'had become so skilful and so accustomed to their tasks, that what appeared at one time to be insurmountable difficulties and hazardous undertakings, had become mere child's play, and was done in those exposed positions as easily as if the men were standing upon the floor of an ordinary workshop'. At certain stages work-platforms could be dispensed with altogether, the men simply carrying out bolting operations by balancing on the bridge itself.

Relations between bosses and Briggers were generally harmonious. Westhofen considered that 'no one need desire to have to do with a more civil and well-behaved lot of men, always ready to oblige, always ready to go where they were told to go, cheerfully obeying orders to change from one place to another, and, above all things, ready to help others in misfortune, not with advice but with hands and purses'. At a time when the union movement was starting to flex its muscles strikes did occur, usually as a result of union committees bringing out men in sympathy with strikers in other industries. Occasionally men would support a fellow workman who had been dismissed for 'an inordinate allowance of the gift of the gab'. Such incidents resulted in 'an immense amount of suffering to scores of his fellow workmen, and more still to their families, and a proportionate increase in the takings of the neighbouring whisky shops'. There were also informal negotiations. When one member of a rivet gang decided not to show, the whole gang would be idle. After negotiation the rivet boys demanded and won a fixed weekly sum of between 20s and 24s a week to be paid by the head of the gang whether they were working or not.

An accident triggered the most significant strike during the construction of the bridge in June 1887. While a moveable stage was being winched up, a girder fouled a piece of timber. Failing to notice this, the men kept forcing the winch until a wheel broke and the stage fell, bringing down a platform on which Briggers were working. Two men and a boy were killed and two others injured. Fifteen hundred workers responded to the usual agitators' call for strike action, demanding a penny an hour more on their wages, representing a 15 to 20 per cent increase, because of the dangerous nature of the work. Extra police were drafted into Queensferry in case of trouble and a rumour swept the town that the strikers were to be replaced by workers from the Tay Bridge, which was nearing completion. William Arrol gave a delegation from the workforce a favourable hearing, promising to ensure that foremen paid out an extra penny an hour for particularly dangerous tasks. Foremen had always had the authority to make such payments but rarely exercised these powers. He also agreed that the proportion of workmen to foremen represented on the Sick and Accident Club committee should be increased. After a week most men had drifted back to work.

During peak periods both day and night shifts operated, although Sunday working was not permitted and the works closed for nearly a week at New Year. In the caissons, for example, work stopped from 6 p.m. on Saturday until midnight on Sunday except for the watchmen who ensured that the compressors kept running at all times. At first the caisson crews worked in eight-hour shifts with eight hours off but later switched to six hours on and six hours off. When working at higher atmospheric pressures their working hours were further reduced to eight hours off between four-hour shifts. The land-based drillers and platers, toiling in the sheds and on the drill roads, had much longer working hours. The day shift ran from 6 a.m. to 5.15 p.m. and the night shift from 5.45 p.m. to 5.45 a.m. The one advantage of the latter was that night-shift workers did not have meal times deducted from their pay.

Lighting the bridge for night working posed particular problems. At first between 1,500 and 2,000 candle-power arc lights were used to illuminate working sites and their approaches. Despite vigorous maintenance they never proved entirely satisfactory with fluctuations in the power and colour of the arcs. They glared and flickered and cast black shadows. Up to 21 lights could go off suddenly when a circuit failed, a serious hazard for men working high up on the scaffolding. According to Westhofen, 'While standing one moment in the dazzling glare of these lights, they were sometimes suddenly called upon to use their eyes in absolute darkness or sit still.' Experiments were made with Lucigen lamps. Using compressed air to produce a spray of oil, stored in 30-gallon tanks, they produced a brighter light than conventional oil lamps. Their drawback was that, especially in high winds, much of the oil could escape unburned, covering the girders and staging with a slippery, slimy film. However good the lighting, at night work-rates were only half those that could be achieved during the day.

The changing pattern of lights on the bridge piers proved a hazard to shipping, especially on dark nights when the men were not working. On one foggy evening the captain of a tug towing a barque downriver mistook a bridge pier for Inchgarvie and headed straight for North Queensferry, which was obscured by mist. He spotted his error in time to back out his boat and slip the tow rope but the barque continued on its way and crashed into the jetty, doing considerable damage to both. Thereafter a lighthouse was erected on Inchgarvie, its flash every five seconds being visible for 12 miles up- and downriver. Standing on a brick pier, the only element of the original design ever built, ironically the lighthouse can be seen as a lasting memorial to Thomas Bouch.

Working conditions were very different depending on the skills and tasks involved. The value of the caisson workers lay in their experience of subsea

Overleaf.

Conference in an Inchgarvie caisson

excavation and their ability to work under pressure, both literally and metaphorically. Local joiners, fitters or labourers also regularly worked in the caissons 'without experiencing any great inconvenience or harm'. Men worked at depths of 63 to 89 feet below high water for an average period of 78 days; the deeper the caisson, the less time the men were allowed to work. It was impossible to whistle in the dense atmosphere and it quickly became hot and humid in the confined space of the 7-foot-high chamber with its arc lights burning. The only bonus was the rare chance to unearth an amethyst or a pearl from the seabed.

Digging out the hard boulder clay was a particularly arduous task even with Arrol's hydraulic spades. The caisson workers tossed the clay into buckets which were winched up by crane, or into skips which were moved to the disposal shafts on trolleys running on rails. Twenty-seven men operating four hydraulic spades could fill a bucket of rubble every five minutes. Speed was also of the essence for the divers who applied quick-setting Roman cement to make good the joints of the caissons. Although not strong, if packed with bags of clay puddle and loaded with stones the cement kept the water out for a time. Divers were also key to lowering the caissons and preparing the foundations for the underwater coffer dams. They shored up the Inchgarvie caissons over the sloping seabed with bags of sand and concrete. The Queensferry north-west caisson flooded on New Year's Day 1884. The divers were later called upon to bolt on emergency plates to allow pumping-out operations to start. After over-vigorous pumping placed too much pressure on the caisson's plates and they burst apart, the divers were summoned again to create a watertight timber frame inside the caisson.

The divers may have operated in teams of three given the photographic evidence of three benches on which they rested before and after dives. Two attendants helped each diver to get kitted up, one to start the air pump and the other to fit the helmet to the diving suit. Dressing a diver involved 17 different stages. The suit was made of vulcanised rubber, lined inside and out with stout twill cloth. When the suit was closed, air was delivered to it from a pump on the surface. The air in the suit also countered the weight of the diver's kit including boots, weights, a knife, and a belt which weighed about 200 pounds. A valve fitted to the breastplate of the diving suit allowed its occupant to adjust the amount of air being pumped in. This gave him control of his buoyancy and could even force him upwards to safety in an emergency. He communicated with the signal man on the surface by a series of coded tugs on a lifeline, the signal man holding one end and the other being tied round the diver's waist. The diver's life depended on the skill and concentration of the signal man who every so often gave a pull on the lifeline. If the diver did not respond immediately with an answering pull on the cord, the signal man brought him to the surface as quickly as possible.

Working in the caissons had its dramatic moments. After arranging a rampart of bags of concrete round a caisson, a diver mistook the bright light of an arc lamp shining in the air chamber for light bouncing off the surface of the water. He swam into the bottom of the caisson, still suspended many feet above the seabed, much to his surprise and that of his fellow workers.

The land-based platers and drillers who cut, bent and shaped the huge flat sheets of steel and drilled holes round their edges for the rivets faced very different challenges. Their work of shaping the plates from wooden guides, created by the

Caisson workers with
hydraulic spades

Above.
Group including two divers in their workwear

Opposite.
Divers and assistants working on the tilted caisson

template cutters from the blueprints, demanded a high degree of accuracy as mistakes were almost impossible to correct at the cutting stage. Although the sheds offered some protection against the weather, it was dirty, sweaty work with the ever-present risk of burns. The men wore canvas aprons known as 'bratties' to protect them against the razor-sharp edges of the plates as they shouldered them between machines. They worked in gangs, each member with his own task but able to perform all the other jobs in the process. Each gang had a leader chosen by the men who negotiated a rate for the job based on the draughtsman's drawing. He might also recruit casual plater's helpers to move the plates around, their pay coming out of the gang's earnings.

Like the platers and drillers, the riveters who hammered in the pieces of metal that held the bridge together were members of the 'black squad', the nickname given to the men whose dirty clothes and blackened faces revealed their lowly status. They worked in gangs of four or sometimes five, their pay being calculated by the number of rivets driven in at the end of each day. Working outside, the riveters had a stark choice in bad weather as no work meant no pay. A boy, often a young teenager, heated the rivets in a furnace or a modified oil-lamp burner until red-hot, removed them with tongs and threw them to the holder-on who placed each rivet into its hole. Sometimes a second teenager conveyed the rivet from furnace to holder-on. The holder-on gave the rivet's head a few hammer blows to lay it up so that it would bed

fairly on the plate. He then held the rivet up with the hammer, placing the head of the hammer against the rivet head while the two men on the other side struck the rivet several rapid blows.

So close was the relationship between erecting and riveting that the rivet teams worked alongside the squads actually putting the pieces into place. Roughly half the riveting was done by hand, the rest using portable, hydraulic riveting machines designed by Arrol. Two members of the rivet team, one positioned on each side of the work, operated a pair of cylinders, one to hold it on, the other to close it. Water pressure from the machine drove each rivet home. The best teams were given the task of hand-riveting the last three bays of each cantilever, the pieces having first been securely bolted together. By the end of construction at least 6.5 million rivets had been driven home. This not only far exceeded the original estimate of 5 million but is probably a conservative figure. An exact count was only kept for Inchgarvie pier which devoured 2.7 million. The figure also excluded the significant number of pieces which were riveted on land prior to erection.

Being a riveter was not an easy life. Other trades tended to look down on the rivet gangs because of the harsh and dirty conditions of their work. In awkward corners a riveter could be bent double, making it difficult to maintain the ringing rhythm of the hammer. Rivets had to be driven in while still red-hot to allow their domed heads to expand and secure the joint as they cooled. Some riveters worked on the moveable stages which consisted of planking supported by two girders hung from a winch by iron ropes. When one section was done the stage was winched to the next section. Each girder was attached to two platforms which were 'walked' up the tower by hydraulic jacks. If conditions were favourable three lifts and the associated riveting could be carried out in eight days. On the vertical columns the men worked in cages, protected by wire netting, with a space of only 4 feet between tube and net. Each cage was deep enough to allow a 16-foot plate to be riveted up without using a hoist. The cages were attached by iron straps to the lifting girders, which crawled like caterpillars up the towers.

A good sense of balance and a head for heights were prerequisites for rivet teams and erectors as they perched ever higher and further out into the river. As resident engineer Mr Cooper laconically described it, 'To provide the necessary standing room for the men when at work on the lower members, light platforms of timber are hung from the tubes by trusses of convenient form . . . men doing the bolting-up either hang onto the parts themselves or stand on planks placed where necessary.'

Rivet teams often included members of the same family. Peter Shearer was born in 1835 on a croft on the Orcadian island of Eday but moved to Edinburgh where he became a tailor and married Margaret Adamson in 1859. They had seven children. Twins John and James worked alongside their brother Peter, who was ten years older, on the bridge; their sister Jessie remembered being allowed to walk out on the cantilevers from the Queensferry side and look down at the water far below her. She treasured a small wooden box with a painted silk inset showing the part of the bridge along which she had ventured. After the end of construction the brothers became brass fitters, the twins learning the trade as apprentices in 1891.

The workforce was not confined to humans. Horses took much of the strain of moving loads of materials from the rail head to the jetties. During the erection of the girders the workmen pushed the wagons along the rails of the approach viaduct to

Riveting the Queensferry cantilever

Opposite top. A rivet team and furnaces

Opposite bottom. Team resting on top of a rivet cage

Opposite. Riveters at work

Below. Horse-drawn bogey on Queensferry Jetty.
The rail viaduct rests on the ground prior to being
raised into position

the work sites on the south side, whereas on the north side Shetland ponies pulled the wagons.

The arrival of visitors regularly broke the monotony of working life. William Ewart Gladstone, four times prime minister, inspected the work while staying with the Earl of Rosebery at neighbouring Dalmeny House. Accompanied by William Arrol, Gladstone appeared particularly anxious that his wife should understand the action of the hydraulic riveting machines. He declined with a regretful smile an invitation to address the workmen who had erected a platform and sought to present a woodman's axe, forged on the premises. After being entertained to tea at Arrol's house, Gladstone entered his carriage and, looking round the crowd of eager faces, said: 'Gentlemen, I wish every success to your great undertaking.'

Privileged visitors were even taken down into the caissons. Philip Phillips described his experience: 'The extraordinary dimensions of these submarine vaults . . . the busy air of activity . . . the noise of the inrushing air . . . the sense of danger entailed by the knowledge that one was for the time being excommunicated from the upper world.'

Members of learned societies, academics attending conferences in Edinburgh, parties of overseas visitors – the bridge became a 'must' for sightseers. Westhofen claimed that visitors to the bridge 'represented a tenth of all people distinguished by

Above. Baker (left) and Arrol (right) with lady visitor

Opposite. Contemporary illustration of
lady visitors on Inchgarvie

rank or by scientific or social attainment . . . As in most other matters, ladies were to the fore, pluckily climbing into every nook and corner where anything interesting might be seen or learned, up the hoists and down the stairs and ladders, and frequently leaving the members of the so-called stronger sex far behind . . . the duties of those called upon to guide the fair visitors were of the most agreeable.'

A typical visit was that paid by the fifth Baron Thurlow, Liberal politician and Lord High Commissioner of the General Assembly of the Church of Scotland, on 24 May 1886. He was accompanied by his wife and a party of 15. They were greeted at the workmen's station by William Arrol, Joseph Phillips and Frederick Cooper, who gave them a guided tour. Having inspected the south span, the party embarked on a steam launch to admire the work on Inchgarvie and the north side. On their return they toured No. 2 Shed, 'where the various machines used in the production of the iron materials for the bridge were examined with much curiosity. The young ladies especially seemed much interested, many of them carrying away iron filings of fantastic shape as souvenirs of the visit.' Miss Phillips then entertained the party to tea before they caught the train back to Edinburgh.

What the Briggers felt as they swarmed over the girders, gawped at from below by boatloads of tourists, is not recorded. Neither is their reaction to their surroundings. Did they have time to rest and admire the view from the summit of the central tower, 360 feet above high-water mark, as Westhofen claimed? 'The broad river itself, with craft of all sorts and sizes, in steam or under sail, running before the wind, cutting across the current on the tack, or lazily drifting on the tide, is always a most impressive spectacle upon which one can gaze for hours with an admiring and untiring eye . . . There have also been many sunrises in early autumn when a hungry man could forget the hour of breakfast and one could not find the heart to chide the worker who would lay down his tools to gaze into the bewildering masses of colour surrounding the light of day.'

The men did make time to celebrate. On his wedding day in 1887 one bridegroom received a gift from his fellow workers of an embroidered outline of the bridge with all its basic measurements, set in a wooden frame, together with an accompanying marriage certificate. By late 1889 leave-taking ceremonies were a regular event. On 27 November, for example, the Inchgarvie workmen presented their manager James Blackburn with a double-barrelled shotgun which no doubt came in useful when he emigrated to farm in Nova Scotia. In his speech Westhofen wished that 'every discharge from it may bring a substantial addition to Mrs Blackburn's larder'. A hooter then summoned the men back to work.

Two days later 25 friends enjoyed supper, speeches and toasts in a local hotel, presenting Mrs Blackburn with a gold bracelet. On 15 January 1890, 'song and sentiment served to pass a happy and harmonious evening' at the Stag Hotel as foreman platelayer David Muirhead bade farewell to his friends, one of many such occasions. By early 1890 the workforce, now made up mainly of painters, labourers and platelayers, had fallen back to 1,000. All the men at No. 2 Shed had been paid off before Christmas, some then later being rehired. By August 1890 numbers were reduced to 80 completing the painting of the insides of the tubes. Thirty-four Briggers were retained permanently: known as ship riggers, they started the never-ending job of painting the bridge. Two men took up residence on Inchgarvie to tend the eight lights that alerted shipping to the bridge. The rest of the Briggers had gone.

CHAPTER 3

BREAKS, BURNS AND BENDS

The Board of Trade inspector was right to be concerned. The Briggers' jobs were not only dirty, deafening and demanding of great physical strength but also downright dangerous especially when working hundreds of feet above the Forth. Many were unused to working at heights, some must have experienced vertigo and a few were undeniably foolhardy. The hazards of such a major operation were legion: from acts of God of wind and weather to the frailties of the human condition such as carelessness or a moment's loss of concentration. The effects ranged from minor bumps and bruises to a 300-foot plunge into the icy waters of the Forth or the agony of caisson disease. Accidents were commonplace given the nature of the work, 430 non-fatal injuries being reported up to November 1887. There must also have been countless unrecorded cases of sprains, pulled muscles, cuts, burns and bruises.

From the start the contractors made arrangements to deal with the situation. A Sick and Accident Club was launched in the summer of 1883 to cover treatment and loss of earnings. Membership was compulsory for Arrol's men, an hour's pay a week being docked from their wages to a maximum of 8d. In return members received medicines and bandages and advice for both themselves and their families. If unable to work they received an allowance in proportion to their contribution, ranging from 9s to 12s a week. Funeral expenses were paid to an upper limit, and for those killed or permanently injured, a grant was made to the family. The Club was funded by an annual contribution of £200 from the contractors, topped up by the profits from the canteen and stores and by donations from visitors. On average 90 men received support from the Club each week until it was wound up at the end of 1889.

In the case of an accident not deemed to be a Brigger's own fault, the contractors normally paid full wages until the man was able to return to work provided that he did not bring a court action against them. In Scotland workers could claim compensation for injuries both under common law and under the Employers' Liability Act 1880. A few accident victims like father-of-six Edward Lafferty took Tancred, Arrol and Co. to court. While Lafferty was helping to erect plate-straightening machinery in 1883, a two-ton iron pillar fell on his left leg, fracturing it in two places. The leg later had to be amputated. The case hung on whether or not the foreman had

We were glad to observe that . . . the safety of the men employed at the great altitude which will be attained as the piers and the cantilevers advance has not been lost sight of.

Quarterly report of
Major-General Hutchinson,
July 1886

Opposite. Queensferry cantilever

instructed Lafferty to help with the machinery rather than leave the site on completion of his job of laying its concrete bed. Tancred, Arrol's lawyers argued that the company had paid Lafferty £1 a week for six months after his accident whereas Lafferty claimed that he had drawn the money from the Sick and Accident Club which he had been forced to join.

In 1888, the year of highest employment on the bridge, the Sick and Accident Club fund stood at £4,546. It was distributed as follows:

Medical expenses	£1,777
Sick allowances	£1,621
Accident allowances	£729
Funeral expenses	£143
Widows' allowance	£176
Donation to Edinburgh Royal Infirmary	£100

The jury found in Lafferty's favour, awarding him £300 in damages on the grounds that he had been inadequately supervised. Eight years later Edward, or Ned as he preferred to be called, was still living in Queensferry working as a labourer. One child had either died or left home but two more children had joined the household, eight-year-old John and baby William. The reason behind the seven-year age gap between

The bridge was served by a flotilla of work boats

the children is a mystery. Perhaps Ned's injuries accounted for it or maybe he spent his windfall on travel.

A similar case was that of Irishman James Boyle of Rosshill, Queensferry, who suffered a fractured skull and internal injuries after being crushed between a large box of cement and the side of a boat from which he was removing stones to a jetty. Five days before Christmas 1884 he was reported as being in 'a critical condition' in the Infirmary, but by the following October he had recovered sufficiently to take Tancred, Arrol to court. He sued for £300 on the grounds that a guy rope should have been provided to guide the box or at least a warning given. He also sued the Sick and Accident Club for £30 due to him, claiming that its committee had denied him information about his case in an attempt to damage his suit against the contractors. The outcome is unknown.

As the case of Glasgow labourer James McGovan reveals, actions were sometimes settled out of court. McGovan had sued for £500 in compensation after a two-ton steel plate fell on him. He argued that the grab lifting the plate was faulty; Tancred, Arrol claimed that the blame lay with McGovan who had not fixed it properly. He was finally 'bought off with a small sum'.

Tancred, Arrol appointed doctors at Dunfermline, North Queensferry, Edinburgh, Leith, Kirkliston and Queensferry to provide medical care for the Briggers and their families. The doctors' duties included looking after the three temporary hospitals at the construction sites: the Queensferry hospital was located conveniently behind the Hawes Inn. An ambulance wagon stood at the ready to take men with serious injuries to Edinburgh Royal Infirmary while those less fortunate were laid to rest in the mortuary on Inchgarvie. Between July 1883 and Christmas 1889, 106 men were admitted to hospital, a proportion of whom subsequently died, and 518 further minor accidents were treated by the doctors on site.

The medical men made around 20,000 visits a year, with Dr Hunter of Queensferry bearing the brunt of a GP's caseload significantly higher than today. In a 12-month period during 1886 and 1887 he made 15,053 visits in Queensferry, Kirkliston, Ratho and around the rural neighbourhood, while Dr Philip made 1,819 visits in Edinburgh and Leith, Dr Menzies 3,480 in North Queensferry and Inverkeithing, and Dr Drysdale 2,147 in Dunfermline.

Even the ambulance house was not immune to accidents. In June 1889 a loaded wagon broke away from the shed and careered round the curve at the top of the incline leading to the Queensferry jetty. Attempts to wedge the wheels having failed, the runaway wagon derailed and toppled over onto the ambulance house, smashing the roof and destroying the interior. Fortunately no one required an ambulance.

James Hunter (1856–1934) was the son of a Dumfriesshire draper and a graduate of Edinburgh University. On 31 July 1884 he married Paulina Glendinning, the daughter of the factor of the Dalmeny estates, at Dalmeny Parish Church. Nine months later almost to the day baby Gwendolina was born at Priory House, Queensferry. In 1887 another Glendinning daughter, Alexandrina, married bridge engineer Wilhelm Westhofen. By the turn of the century the Hunters and their five children were living at St Catherine's, Linlithgow where Dr Hunter remained until his death.

Management tended to blame the men for the doctors' heavy caseloads. Westhofen believed that three-quarters of accidents were avoidable, regarding the workmen as indifferent to or careless of the risks to themselves and others. If, however, an accident occurred they would go to the aid of their mates without thought of any danger to themselves. Philip Phillips also blamed the Briggers. 'The men by constant employment at these heights seem to get used and hardened to the work, the result of which is carelessness.'

Arrol took a less judgemental view considering that accidents often resulted 'from some very simple thing – walking across some planks from one place to another, instead of taking the proper way, or standing carelessly on a girder and slipping off.' Working day after day suspended on platforms high above the river, it was inevitable that some men simply forgot where they were. It must have been all too easy to step back to check that a rivet was holding firm or even just to admire one's handiwork. While carrying planks to cover up a hole in the scaffold, one worker walked backwards into the hole. Boys leapt from plank to plank like agile monkeys heedless of what lay below.

Standing on the edge

Management blamed drink as the root cause of many accidents. Arrol claimed that he regularly saw drunks hanging around the works gates, although no one was permitted on site while drunk. In a speech, albeit delivered at the Tay Bridge in 1886, he compared the record of his two workforces in terms of intemperance. 'At the Forth Bridge we are very much troubled with it; at the Tay Bridge, we are not troubled with it at all.' He singled out the British caisson workers as particularly prone to temptation. Short shifts, easy work and feeling 'a little elevated' because of the change in air pressure often resulted in them spending their six to eight hours of free time between shifts in the pub, thus rendering them unfit to return to work. By comparison the Italian caisson workers rested on their beds until the start of their next shift. Arrol, who claimed not to be a teetotaller although he did not drink, considered that intemperance was the greatest management challenge he faced. In return the *Dunfermline Journal* pointed the finger of blame at Arrol himself as it was his company which had approved, or at least turned a blind eye to, the sale of cheap

Most briggers were confident working at heights

drink in the works canteens. 'At the outset the contractors sowed the wind, and now they are reaping the whirlwind.'

In turn Arrol pointed his finger at the Hawes Inn. Acknowledging the widespread criticism of his previous remarks, Arrol nonetheless returned to his theme that 'the further a public house was from a public works, the better' in a speech at the Thistle Hall, Dundee, in 1887. There is no doubt that many Briggers were hard drinkers. On pay day the Hawes Inn lined up 200 pints at a time on the bar and had to introduce drinking in shifts. Arrol estimated that 2.5 per cent of the men's wages ended up on the counter each week. Westhofen thought that Queensferry had far too many public houses.

The men had to work in all weathers with temperatures ranging from 20°F to 85°F. Time was lost to bad weather, work being able to proceed on 22 or 23 days a month on average. The prevailing wind came from the south-west, followed by north-easterlies which whipped up the waves on the estuary and were trying on the

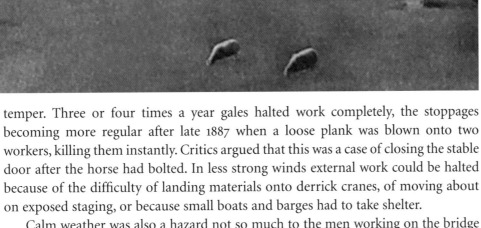

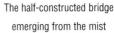

The half-constructed bridge
emerging from the mist

temper. Three or four times a year gales halted work completely, the stoppages becoming more regular after late 1887 when a loose plank was blown onto two workers, killing them instantly. Critics argued that this was a case of closing the stable door after the horse had bolted. In less strong winds external work could be halted because of the difficulty of landing materials onto derrick cranes, of moving about on exposed staging, or because small boats and barges had to take shelter.

Calm weather was also a hazard not so much to the men working on the bridge but to the small sailing boats that could drift helplessly around it if the wind dropped suddenly during a strong ebb tide. The work boats and barges were instructed to look out for small boats in trouble and tow them into mid-channel. Three timber booms were moored to the west of Inchgarvie to fend boats off the island's iron staging and vice versa. These booms saved countless sailing boats from shipwreck.

Early morning rain sometimes drenched the men, who downed tools and went home for the day despite the weather improving in an hour or so. Westhofen was

puzzled: 'No power of persuasion was great enough to bring them back again even if the weather turned fine and continued so for the rest of the day – a curious fact not easy of explanation.' Although there was very little snow over the seven winters of construction, frosty weather regularly resulted in stoppages when the water used as a fluid for hydraulic equipment froze. One man working at 100 feet in bitterly cold conditions relied only on a rope grasped in one hand to keep his balance. When his hand became numb, his grip loosened and he plunged into the water but survived to tell the tale.

The Forth was subject to sudden sea fogs, known as haars, which came rolling up the river especially in spring and early summer. Sometimes the top of the bridge would be in sunshine while the lower stages were shrouded in thick white mist; at other times the situation was reversed with the towers lost in the cloud. Occasionally a bank of mist would hover over the middle of the bridge with azure sky above and sparkling sea below. Helmsmen of the steamers shuttling several hundred Briggers to and from work had to steer their way with extreme caution through busy shipping lanes with the additional hazard of the bridge itself looming through the shifting haar.

Other than halting work altogether, there was only a limited amount that management could do to protect the workforce against the weather. The shelters on the bridge which provided hot food and lashings of tea helped the men to avoid a drenching and the resultant colds and chills during unexpected squalls and hail showers. Clothing offered limited protection; in the 1880s lightweight weatherproofs did not exist. Many Briggers wore their oldest, much-patched trousers and jackets, with a cardigan or muffler for warmth and a battered 'bunnet' to protect the head from wind and rain. The contractors gave the men working on the foundations free waterproofs and boots. During the erection phase they provided thick woollen jackets, overalls and waterproof suits which were heavy and prone to leak. A nominal charge was made to encourage the men to treat the clothing with respect but this was often waived for 'careful and deserving men'.

Water was an ever-present danger, especially given that many Briggers were unable to swim. Accidents could happen even in relatively shallow water. While working at the Queensferry jetty, for example, a painter's lad toppled into only 20 feet of water. On being rescued he was found to be 'much cut about the head and bruised about the body', his injuries probably being sustained from the rocks below. Deeper water sometimes acted as a blanket helping to break a fall. In 1889, 12-year-old John Nicol from Dunfermline fell 80 feet from the Inchgarvie staging. On examination by Dr Hunter he was declared to have suffered nothing worse than a wetting.

Some individuals survived falls from even greater heights. One such born survivor was 17-year-old Thomas Dow from Leith, a rivet heater working on the Queensferry cantilever. About ten o'clock on the morning of 22 January 1889 he slipped from the staging and fell over 100 feet into the icy water. While onlookers summoned a boat to the rescue, driller John Robertson, already a hero of several rescues, ran along the jetty and jumped in. The boat reached the stricken teenager first and hauled him on board. He was carried to the ambulance where the superintendent, Sergeant Manson 'at once applied the proper restoratives' until Dr Hunter arrived. Despite his experience Thomas was able to answer the doctor's questions about his state of health. After resting until four o'clock he was removed to the Infirmary to recover from 'a severe shock to his system' and suspected internal injuries. Robertson survived his heroic dip unharmed.

One of the most remarkable survivors was rigger Alfred Lawson, who was seized by an attack of cramp in his arm and lost his grip while climbing a rope over 100 feet above the Forth. Despite taking some time to reappear on the surface of the water, he was none the worse for his adventure. Humans were not alone in having miraculous escapes. In 1890 Mr Ramage, a lifelinesman on *Defiance*, the Forth Bridge Company's

One wrong move?

boat used to transport divers, saved the life of a squirrel which had slipped while scampering along the Inchgarvie cantilever. A fortnight later it was Mr Ramage's turn to be rescued when he fell overboard. He was none the worse for his ducking.

The contractors put the appropriate measures in place. A rowing boat manned by two expert watermen was attached to each cantilever for emergencies. The safety boats saved at least eight lives as well as 8,000 caps and other garments blown off the bridge. A steamboat which worked at night towing barges with their holds full of water for storage on Inchgarvie also acted as an emergency vessel. Some of the workmen proved to be strong swimmers, diving in after their mates when they plunged from the scaffolding. One such hero was Queensferry labourer Peter Carroll, who dived into the Forth after a loose plank catapulted a fellow Brigger into the water 200 feet below. Sadly, on this occasion his rescue attempt was defeated by the currents.

Air was a more unexpected and less well-understood elemental hazard. Although the contractors perceived the caisson operations to be the greatest threat to life and limb, their concern was focused on the lowering of the giant cylinders to the seabed. During this operation only the buoyancy of the caissons prevented them from descending suddenly and smothering the men, a major cause of fatalities in similar projects. The risks were highest at low water when the men were often taken off the job and the air pressure reduced to allow the caissons to descend at their own speed. On one occasion a caisson suddenly fell 7 feet, causing the air chamber and part of the ascending shaft to fill with mud and silt. A more serious incident resulted in four men being injured. They were filling a tube with concrete in order to lower it to the bottom when the men below opened a valve to let air into one of the service locks. Because the door at the top was closed, air rushed up into the first chamber with such force that it shattered the fittings and the four men were buried.

The more serious, but hidden, risk, however, was caisson disease, or 'the bends'. At the time the condition, which is due to gas bubbles being created and then released into the blood and tissues when the air pressure changes abruptly, was not widely understood. Observers noted that caisson workers suffered strange and alarming symptoms, notably a tendency to stagger and agonising cramps in their joints shortly after coming off shift. In 1859 a worker died from caisson disease while working on Brunel's Royal Britannia Bridge in Cornwall. During the building of the Eads Bridge (1867–74) over the Mississippi, 15 workers died and 79 suffered the severe effects of caisson disease. Owing to a bitter dispute between engineers James Eads and Washington Roebling, designer of New York's Brooklyn Bridge (1870–83), lessons learned from the Eads Bridge were not passed on. Three Brooklyn Bridge workers died of caisson disease and 15 per cent experienced a degree of paralysis. Roebling, a regular visitor to the caissons, was afflicted by the effects of the bends for the rest of his life. Workers on the Brooklyn Bridge nicknamed the excruciating condition where a man could double up in pain 'the Grecian bends', after the posture adopted by fashionable women of the era with tight corsets and big bustles.

The Forth Bridge contractors drew on the American experience, from providing a hearty meal an hour before the start of a shift and hot drinks at the end to ensuring that workers ascended gradually to the surface. The caisson workers protected

Opposite top.
A brigger above entrance to caisson air lock

Opposite below.
Workers at air lock of materials shaft

themselves by their own fashion accessory, arm and ankle bands of silver and zinc scales. Management put forward a catalogue of possible reasons: wet feet, hard work, excessive sweating, a too sudden change from a heated atmosphere to a biting east wind, lack of suitable clothing, malnutrition, prior ill-health and staying in the caissons too long. Westhofen observed that returning to the caisson often alleviated the symptoms:'Many of those afflicted with the disorder spent the greater part of Saturday afternoon and Sunday under pressure, and only came out when absolutely obliged to do so.' He attributed the root cause as usual to drink: claiming that even the most experienced foreign hands suffered when they had 'been making too free with the whisky overnight'. It may have been, however, that some of the workers reeling out of the Hawes Inn were suffering from the bends. Arrol himself paid for spending too long in the caissons by partial deafness for the rest of his life.

Dr Hunter became fascinated by caisson disease, producing a thesis – 'Compressed Air – Its Physiological and Pathological Effects' in 1887. He faced several handicaps in conducting his research. 'The inaccessible position of the working chambers of the caissons, the fact that the workmen were foreigners speaking a different language and having prejudices against being made the subjects of scientific observations; the constant noise of machinery, my other professional duties only leaving Night as the time available for this special work; – these, taken with the interruption caused by an outbreak of small-pox in my practice, were some of the difficulties I encountered.' Hunter visited the caissons on several occasions, recommending eating a biscuit to stimulate saliva and relieve the excruciating earache experienced by many caisson workers after the stopcock was turned to release compressed air into the airlock before entering the chamber proper.

Hunter was a born experimenter. He observed the effects of compression on his own body, from difficulty in whistling unless kneeling with his head facing downwards to experiencing a headache for several days after visiting a caisson. He persuaded a group of friends to enter No. 4 Caisson to record changes in their pulse rates and on another occasion took a white rabbit with him whose pink ears and eyes turned paler during the two hours it spent in the caisson. He noticed that experienced workers tended to be thin and anaemic despite increased appetites. Even bribery, however, failed to persuade the foreign workers to allow him to collect their urine over 24 hours. He studied the effects of different workplace situations. In his view the men were most at risk when removing soft silt at the start of operations, because of dampness and decaying animal matter, and again at the end when concreting the caissons on the sea bottom because of the accumulation of stale air and carbonic acid gas.

Ironically the first caisson worker that Dr Hunter treated was not suffering from the disease. A young Frenchman contracted typhoid after the sinking of the first caisson, his symptoms being attributed to a shoal of decomposing fish trapped underneath it. More likely, it was caused by sea water contaminated by sewage. Among the first workers to show signs of caisson disease were a group of young Belgians who, after walking back to their huts on a frosty day at the end of a shift, suffered agonising joint pains, especially in their knees. One worker described the pain as being 'as if a knife were thrust into the joint and turned backwards and forwards until its contents had been removed'. Another likened it to his leg and thigh being stretched as far apart as possible and then snapped into place again. The

recommended treatment was bed rest, flannel clothing and a twice-daily dose of Ipecac with Dover's powder. Ipecac, a bitter extract of a tropical root, acted as an emetic to ensure that the patient did not ingest too much of the opium contained in Dover's Powder. Invented by the physician and adventurer Thomas Dover in 1732 and available over the counter in chemists, the powder remained a popular remedy for pain relief until its addictive qualities were recognised in the 1900s. Usually after two or three days the men were deemed fit to return to work.

Stomach cramp accompanied by vomiting and giddiness was another symptom of caisson disease. One victim was a young Belgian called Rayen. After a hearty meal and a change of clothes in preparation for a trip to Queensferry, he collapsed in his room on Inchgarvie. On being summoned, Dr Hunter discovered that Rayen had recently taken two days off work with joint pains. After a week of lying still, consuming milk and ice and being dosed with indigestion remedies, Rayen was pronounced fit although he decided to return to Belgium rather than face the caisson again. Another worker had to be sent home after going insane.

One young Italian nearly died of pneumonia after coughing up a massive amount of blood as he lay resting on the deck of the *Hougoumont* at the end of his shift. It turned out that he had suffered a lung haemorrhage some years previously. His fitness for work might have been queried if the Forth Bridge contractors had provided regular medical examinations as was the case with caisson workers building the Eads Bridge. Dr Hunter commented, 'Owing to the contract for the caisson sinking being in the hands of foreigners with previous experience of similar work, no such precaution was adopted.'

Some men experienced severe giddiness, on its own or accompanied by other symptoms. Fritz Ingorvik, a tall, thin German aged 48, had been working on the bridge as a labourer for some months before volunteering for caisson work, despite having no previous experience, on the grounds that the pay was better. After his first shift on No. 1 Caisson he walked a mile to his lodging in Queensferry where he collapsed with severe giddiness and headache. 'He reeled and staggered worse than a drunken man.' Even after months of treatment he veered 3 feet to the left when attempting to walk across Dr Hunter's surgery in a straight line.

In all Dr Hunter treated a sixth of the Forth Bridge caisson workers without, in his view, any fatalities or permanent disabilities. At a time when the mechanism behind the disease was not fully understood Dr Hunter preferred the idea that the differing symptoms related to different biological effects of decompression rather than backing any one of the theories put forward by eminent medical men of the time. In 1907 Scottish doctor J.S. Haldane came up with a means of largely preventing the bends by staged decompression, producing the first statistical tables of safe diving depths.

Both on land and on the bridge, work sites were a chaos of moving parts as cranes swung, wagons tipped and wheels spun. James Blennie was the victim of one of the first accidents reported in the press when on 20 June 1884 he was struck on the back by the chain of a crane. His spinal injuries were sufficiently severe for him to be moved to the Infirmary three days later. On 26 July carter Alexander McDougal ended up in the same place with a compound fracture of his left leg. He had been shunting wagons when his horse had suddenly turned and knocked him down, the wheels of two of the wagons then running over his leg. Soon 'Accident at Forth

Bridge Works' became a standard headline of the *Dunfermline Journal*.

The works were dangerous places. Nineteen-year-old Donald McDonald had plenty time to rue his rush to breakfast – accidents often happened in the hour before meal breaks – when his flapping jacket caught in the teeth of the wheels of a drilling machine. The force swung him round, resulting in other parts of his clothing also becoming entangled. Before he could be rescued his clothes were in tatters and a wheel had passed over the fleshy part of one of his arms.

A particularly harrowing case was that of labourer John Tansay, who fell into a 9-foot-deep saw pit on 23 November 1886. His cries went unheeded until the next morning when he was carted off to the Infirmary with severe spinal injuries.

Machines were accidents waiting to happen. Henry Loudon had a lifelong reminder of his time as a Brigger after losing two fingers of his right hand when it was crushed between two cogs of a punch machine that he was oiling. Driller John Moffat broke several ribs after his clothing became caught in a vertical boring machine. Remarkably, however, there was only one fatality: when in mid December 1885 John McDade was hit by a plate at No. 2 Shed.

Their unexpected swinging and imperfectly secured loads made cranes a particular hazard both on and offshore. An early victim was works labourer John Dalgleish, who suffered a fractured skull when a log slipped its chains and knocked him into a 20-foot-deep coffer dam. Fourteen-year-old James Rennie fractured his right leg in several places when one of the Inchgarvie cranes toppled onto him. In September 1886 two young men, John Weir from Dunfermline and George Smith from North Queensferry, were admitted to the Infirmary with fractured thighs and bruising after an accident with a crane. About four o'clock on the same July day in 1887 as two men died and four were injured when staging on the bridge collapsed, Hugh Foley was employed on Inchgarvie loading coal. A rope with an attached hook, which he was using to guide a crane bucket, slipped and he overbalanced, falling 28

feet and breaking his arm.

The caissons took their toll not only during their operation but while they were being built in 1885. William Forrest, for example, was manoeuvring a box of bricks by crane to the top of a caisson when the box swung awkwardly and crushed him between box and caisson. Several workers were seriously injured when they lost their balance and fell inside a caisson.

During the main construction phase the risks increased dramatically and new solutions had to be found. Nearly 20 cages hoisted men and materials to the work platforms. Such lifts were potentially lethal as the ropes that hauled them up and down were apt to fray and snap due to the constant friction. Acutely aware of the problem, the engineers came up with the solution of using steel-wire ropes. Despite operating night and day none of the ropes gave way, although some had to be replaced on showing signs of wear and tear. According to one estimate, replacing hemp ropes with wire cables reduced the accident rate by half.

As work progressed hundreds of tons of equipment – including cranes, temporary girders, winches, steam boilers, rivet furnaces and riveting machines – miles of wire ropes and acres of timber staging and gangways were suspended from the cantilevers. Management was particularly conscious of the risks associated with the work platforms which regularly had to be moved, often with the men still on them, during the building of the cantilevers and central towers. The platforms were strung one above the other like flats in a multi-storey block, each one home to scores of Briggers. They were littered with the detritus of the working day. A nut or bolt, much more so a hammer, rivet tool or bucket falling from a height was enough to kill or maim a man on the platform below.

Despite the resident engineer's view that the scaffolds and work platforms meant that 'men should be able to work at any height in as great security and comfort as if on the level of the ground', it is possible that some Briggers forgot momentarily but fatally that they were working hundreds of feet up. The scaffolding stages were barely wide enough to take the workmen, their tools and their clutter. The men regularly dropped or threw to each other hammers, chisels and bits of timber. Many of these ended up over the side, tumbling down three or four tiers of staging below. When one workman dropped his spanner it fell 300 feet, passing through a four-inch plank. Another spanner whistled through the air before entering a Brigger's waistcoat and exiting through his trouser leg. Apart from ripped clothes he was unscathed. Gangs of workmen were specially employed to keep the site tidy and regular warnings were issued, largely to no avail. According to Westhofen, 'It needed the sight of a wounded and mangled fellow creature, or his bloody corpse, to bring home to them the seriousness of the situation.' Wire netting was stretched on frames above and under work areas to catch falling tools. Handrails were fitted to scaffolding where men were moving around and edges were attached to suspended platforms to stop bolts and rivets falling off.

Given the nature of their work, rivet boys were particularly vulnerable. On 8 September 1884 an unnamed 14-year-old living in North Queensferry fell from a scaffold, fracturing his thigh. Less than three months later the ambulance was on its way to the Infirmary again, this time with a 16-year-old Queensferry rivet boy James Todd, who had fractured his right leg when he fell 40 feet from the top of a caisson. The higher the bridge rose the riskier the job became. In January 1887 rivet heater

John Evans fell from the top of one of the main tubes, ending up in the Infirmary with concussion.

The wooden staging on the cantilevers and towers also presented a fire risk given the volume of riveting. At first gangs heated their rivets on portable furnaces but these were soon deemed too dangerous. Not only did the furnaces and the coal they consumed add additional weight to the platforms but their hot ashes could easily spark a fire. After experimenting with various devices the engineers came up with the solution of a portable, iron, oil-burning furnace lined with bricks. It was small and simple enough for rivet boys to operate, lighting it with a piece of burning rag soaked in oil. The boys used to connect the furnaces to the compressed air pipes, thus saving themselves the trouble of having to work the bellows to achieve a constant blast. They could turn out 200, and in emergencies 250, rivets an hour in return for feeding the furnace two gallons of oil. As well as reducing the risk of fire, the furnaces also proved exceptionally easy to move around by crane.

Fires did occasionally break out. In October 1888 a workman discovered that the staging inside one of the cantilever tubes was alight. Two men were dispatched down the tube to put out the fire but had to be rescued with ropes when overcome by smoke. They soon recovered 'on the application of the usual restoratives'. Four months later fire broke out during the night at the top of the Inchgarvie tower. A south-westerly gale threatened to fan the flames and damage the structure below. The gale made reaching the island difficult, much less the climb up the super-structure to close the joints, which had been broken to avoid frost damage, and allow

Digging out the north approach railway

Opposite.
The railway cuttings took their toll in life and limb

the hydraulic pumps to spread water on the blaze. In December 1889 a spark from a naphtha lamp resulted in the shelter for the horses at the head of the Hawes jetty catching fire. While pushing the shelter into the sea to douse the flames, the men damaged two granite blocks destined for one of the cantilever piers; the fate of the horses was unrecorded.

Chemicals, too, posed dangers, one Brigger on Inchgarvie suffering extensive burns to his hands and face when his naphtha lamp exploded. They also affected the health of the Briggers more insidiously. The men scraped and brushed with steel tools all the plates, bars and angles for the superstructure as soon as they left the works. They then coated them with boiled linseed oil. Once the components were in place on the bridge, painters gave them two coats of red lead paint followed by two coats of iron oxide paint. They also applied one coat of red lead and two of white to the insides of the tubes. Given their exposure to paint fumes, some must have suffered the severe digestive problems associated with 'painter's colic'.

One of the most significant causes of accidents was the building of the approach railways, especially the tunnel between North Queensferry and Inverkeithing. During 1889 reports appeared in the local press with a regularity that numbed readers' sensitivities: 'right foot badly crushed by a wagon'; 'fracture of the right leg by being struck by a stone'; 'right leg severely fractured by a fall of earth'; 'one of his legs broken in two places by a piece of rock falling upon him'; 'skull fractured and his

left shoulder injured, through being knocked down by one of the ribs of the arch giving way'; a fractured skull, a dislocated shoulder, and other internal injuries by a fall of stone'; 'somewhat seriously injured by a stone falling upon him while engaged in a cutting'; 'crushed against the side of the cutting by a wagon'. The casualty department of Edinburgh's Royal Infirmary was rarely busier with industrial accidents.

Even commuting could be risky, especially if one had spent Saturday evening in the pub. Rushing to catch the Port Edgar steamer on his way home to Dunfermline, labourer Robert Robertson tripped on a rail and injured his hand. After having it dressed by the stationmaster's wife, he decided to walk back to Queensferry. Feeling the need to lie down for a bit, he jumped over what he assumed was a field dyke but in fact was the parapet wall of a bridge over the railway line. He was discovered next morning lying on the railway line suffering from injuries to his arm and spine. Railway navvy Patrick Campbell caught a workmen's train at Queensferry but forgot that it did not stop at Dalmeny where he was working. So he decided to jump off as the train passed through Dalmeny with catastrophic results: 'It is expected that both legs will have to be amputated.'

The doctors also had to contend with the unexpected. Early in 1886 smallpox broke out in a workmen's hut and quickly spread to a densely populated and insanitary area of town. The burghers of Queensferry had been tardy in taking on their responsibilities for street cleaning and sewerage. Belatedly the town council ordered

Briggers at drill roads

the streets to be disinfected and all open rubbish stores and ash pits to be closed. The *Hougoumont* was towed to Port Edgar and fitted out as an isolation hospital. By early March Dr Hunter was caring for 24 *Hougoumont* patients, of whom ten were convalescent: five had already been discharged and three were being treated at home. By the end of the epidemic, despite stringent controls and an emergency tax on landlords and home owners, three or four patients ended up in the damp and overcrowded churchyard. An English firm bought the *Hougoumont* in 1889 and towed her away.

No one denied the potentially lethal nature of the working environment. Dozens of Briggers were left with a lifelong reminder of their time on the bridge. Labourer Thomas Young of North Queensferry lost two toes, sliced off by a falling stone. Riveter James Simpson had an eye knocked out by a spark from a piece of metal. Michael Kelly lost a leg when a stone crushed it during tunnelling operations for the railway approach from Inverkeithing. Arrol was known as a good employer and mindful of his men. Westhofen's view was that 'Every care was taken and no expense was spared to make good and secure staging for the workmen, and to construct gangways and roomy staircases for all places where work was carried on.'

While noting the increased number of accidents in 1885, *The Scotsman* considered that 'it would be difficult to suggest that these were in any way due to the quality of the plant, or to the want of adequate precautions taken by the contractors for the safety of the men'. Any conclusion as to how many of these accidents were avoidable has also to be set in the context of the knowledge and technology available at a time when health and safety legislation was in its infancy. What caused controversy both at the time and ever since was the degree to which the contractors seemed to regard the Briggers' lives and limbs as an acceptable human cost in return for the economic and social benefits of bridging the Forth.

CHAPTER 4
CRUSHED, SLIPPED OR DROWNED

On Tuesday 27 November 1883 Thomas Joseph Harris earned a footnote in history as the Forth Bridge's first fatality. The 16-year-old storekeeper from Queensferry was last seen, as dusk fell, standing well out on the nearly completed 2,100-foot-long jetty where he had been working. There was no one to witness his fall or hear his screams as most workmen had gone home for the night. In the coming days local fishermen and the crew of the work boats dragged the river in vain. Thomas's grieving father, bridge works foreman Frederick Harris, finally registered the death on 20 December. Another month passed before Thomas's body was discovered, washed up on the shore 200 yards east of Barnbogle Castle, the recently restored retreat and library of the fifth Earl of Rosebery. Thomas's body was so decomposed that it was identifiable only by its clothing.

The next death, in the early hours of 30 June 1884, was very different. John Fairless, a 46-year-old master mariner from Middlesborough, was captain of the steamship SS *Diana* which had sailed from Arbroath two days earlier with a cargo of flagstones for the cantilever foundations. The day's unloading from the hold onto the pier at Inchgarvie began at 3 a.m. Half an hour later, possibly roused by the noise, Captain Fairless was levering himself from his bunk when there was a tremendous crash. The cable of the steam crane had snapped under the weight of a stone, weighing nearly two tons, which careered through the cabin roof and fell on Captain Fairless's head. He died shortly thereafter. The same day his son, who was with him on the ship, registered his death, the first of many to be certified by Queensferry GP Dr Hunter.

Two other deaths occurred that year as the groundwork for building the bridge gathered pace. In July a middle-aged, unmarried labourer, Robert Adamson, was crushed between two railway wagons at North Queensferry, dying from 'Compression of Thoracic Viscera & Shock'. Four months later, fitter Kenneth McDonald from Inverkeithing fractured his skull after falling 40 feet from scaffolding to the bottom of No. 4 Caisson, which was in the course of erection at the Forth Bridge Works. He survived for the best part of two days before succumbing to his injuries, too ill to be moved from the adjacent ambulance house.

It is impossible to carry out a gigantic work without paying for it, not merely in money but in men's lives.

Benjamin Baker

Replacing bolts with shiny new rivets

The caissons claimed two victims. Although Benjamin Baker boasted in a speech to the Royal Society in 1887 that 'happily, no serious accident happened' during the sinking of the caissons, his memory must have been playing him tricks or his definition of an accident rather narrow. Two men died after the Queensferry north-west caisson ruptured at 6.45 p.m. on 25 March 1885 during the long-drawn-out attempt to right it after it toppled over. Pumping operations had been carried out all the previous day with the result that the water inside the caisson had been reduced by about 12 feet below the level of the water outside. The external pressure caused the recently installed metal plates to give way. Seven men were working inside the caisson when a great rush of water engulfed them. The force of this 'whirlpool of death' was sufficient to smash heavy wooden beams into matchsticks. Five men floated to the top of the caisson, escaping with varying degrees of injury. Boiler maker John Gibson from Inverkeithing and his friend Alexander Ferguson, a joiner from Queensferry, were less fortunate. They failed to reach the surface and drowned, their ascent thought to have been hampered by the same debris that later was to prove a hindrance to the divers searching for their bodies. Both victims were married men in their thirties, Gibson leaving a family of five children. The tragedy could have been even greater as during the day shift, which had ended two hours earlier, 70 men were working in the caisson.

The death toll inexorably mounted over the following years, peaking in 1887 with 22 fatalities as the central towers rose and the cantilevers inched outwards and upwards. When project timescales started slipping, more and more men were thrown at the job.

From late 1886 the catalogue of deaths becomes remorselessly routine. Rivet heater James Hennan (25) missed his footing on top of the girders . . . Labourer Angus Paterson (21) died in the Infirmary of his injuries after an iron plate, which he and six other workmen were lifting, slipped and struck him on the chest . . . Labourer Thomas Birrell (51) lost his balance when the cage inside which he was working lurched, falling 150 feet to instant death . . . Alexander Steel (26) was hit on the head by an iron plate as it swung upwards on a hoist . . . Patrick McGarry (16) overbalanced and fell 50 feet head first . . . Plater Peter Burns McCowat (49) was taking orders from his foreman on Inchgarvie when he was hit on the head by a falling plank from the cantilever above. Rigger John Anderson (30), alias Eugen Julius Weise from Germany, fell 200 feet from a platform and fatally fractured his skull . . .

Why Eugen Julius Weise became John Anderson remains a puzzle – just one of many faced by the research team in ascertaining the actual number of bridge deaths. On 30 March 1887 *The Scotsman* reported that 'John Anderson (30), rigger, fell from the top platform of the vertical columns at the end of the jetty on the south side (200 feet) and was killed, the skull being fractured'. A search of death certificates failed to find him. What it did reveal was that 20-year-old Eugen Julius Weise, a rigger from Königsberg, Germany, died from identical injuries on the same day although many of the details from the death certificate are missing. The team concluded that Anderson and Weise were the same man. Did the reporter simply use the wrong name or is there a story about the alias still to be uncovered?

The accident rate accelerated as winter 1887 approached. Hit on the chest by a bar which sprung from its position as he was loosening a bolt, plater James Munday (20) was catapulted onto the staging below . . . Labourer Frank McLean stepped off his

crane, missed his footing and fell 150 feet fracturing his neck, skull and jaw . . . Holder-on Montague McNab (47) and carpenter William Rendall (30) were hit by flying planking during a gale . . . Robert Latto (21) was crushed when a bolt snapped as he was trying to move his riveting machine with a block and tackle. The machine swung back, jamming him between it and the cantilever . . . Labourer Joseph Courtney (23) was throwing coal into a crane's furnace 350 feet up on a cantilever when he overbalanced and fell over a handrail on which he had been resting his foot. He landed on his head on the internal viaduct, 200 feet below.

The year 1888 started propitiously with no serious accidents until 19 March when nightwatchman Robert Reid was reported missing; his badly decomposed body was found two months later. On 23 March an unprecedented accident occurred when a two-ton crane, weighed down by a steel plate swinging from its jib, gave way on a

Deaths by year

1883	1
1884	3
1885	5
1886	7
1887	22
1888	17
1889	10
1890	3
1891	4
1892	1

viaduct. As it plummeted into the sea it took James MacCallum with it. He was dead long before he hit the water, encountering some projecting ironwork: 'His skull was split open, the poor fellow's brains being bespattered about.' The accident was newsworthy as much for the 'miraculous escapes' of the three men on the platform beside McCallum and the three Briggers working underneath the viaduct. Despite falling 60 feet, William Valentine was able to swim to Inchgarvie and afterwards walk home with little more than a bruised shoulder. Until the facts were known people in both the Queensferries feared a much greater tragedy and the workforce was sufficiently rattled that some Briggers downed tools for the day.

Over the summer the litany of deaths continued with relentless regularity: William Cunningham, a stoker who died on board the steamboat *Invergowrie*; Hume Smith, a mechanical engineer's labourer who made it to the Infirmary but died of his injuries; Thomas Roberts and William Fairley, variously described as boatmen, fishermen or nightwatchmen; John Curran, William Brown and William Wright, who all fell to instant death from the top levels of the bridge in under a month. The autumn was little better with one death in September and four fatalities in October. Thereafter there were no deaths until the end of April 1889 when joiner's labourer Thomas Vennert died in the Infirmary of shock and blood loss after his leg was amputated at the thigh.

Even as construction drew to a close and the workforce fell to under 1,600, fatalities continued. The twenty-seventh quarterly report of inspection by Major-General Hutchinson and Major Marindin to the railway department of the Board of Trade regretted that 'even with this reduced number of workmen, there have been four fatal accidents'. It attributed the death of riveter John Aitken (17) on 13 September 1889 'to carelessness in the fixing of a stage upon which he was working'. Two deaths less than a week apart in September were 'due to the falling of pieces of wood from a height, which should not occur if proper care be taken'. The report considered the last case to be a genuine accident when a rigger lost his balance and fell into the water.

In 1890 Wilhelm Westhofen wrote: 'Of accidents between July, 1883 and Christmas, 1889 when the Club ceased to exist, there had occurred 57 fatalities.' His figure had never been challenged until members of The Forth Bridge Memorial Committee, set up in 2005 to erect a fitting monument to the Briggers, posed the question: 'What are the names of the men who died during the construction of the bridge?' For many years a number of Queensferry residents had passionately felt that it was more than time to recognise the courage of the men, their counterparts who died building the Forth Road Bridge having long since been fittingly acknowledged.

A team of local historians set out to answer the question. It seemed deceptively easy: find the records of the Sick and Accident Club and the names will be revealed. Despite an exhaustive search, however, they could find no trace of the records other than in newspaper reports. The team scoured the national and local press over the seven-year period in question. In order to reduce this Herculean task they enlisted the help of professional genealogists Ian Stewart and Sheila Hay, who worked through the death records stored at New Register House in Edinburgh. They produced a steady stream of possible names and dates which could be matched to contemporary newspaper accounts.

By trawling police court and hospital archives and other sources the researchers gathered a mass of additional detail. Often tantalisingly and frustratingly

fragmented, this shed a little more light about the lives and deaths of the Briggers, the conditions under which they worked, and the daily routine and dramas of the communities on each side of the Forth. Seeking answers to specific questions opened up other avenues of research. Was Queensferry hit by a crime wave given the Briggers' reputation among contemporary commentators for hard drinking and wild behaviour? To what extent was drink the root cause of accidents or no more than a natural escape route after a hard shift on the freezing cantilevers?

Even matching names and ages quoted by different sources required significant detective work. In an era when records were handwritten and newspaper reports not always accurate, several versions of the same name or event could emerge. It is under-standable that the clerk might have difficulty with the name of Nicolia Liberale or

Nicholi Librali, the 40-year-old Italian who died on the jetty on 11 March 1885. Similarly it is explicable that little is known of his personal circumstances other than that he was married and was residing in Queensferry, given that he was probably an itinerant worker and possibly spoke no English. His death certificate was signed by Inspector Thomas Rawlins, suggesting that Liberale met a lonely end. What is unusual but by no means unique is that the cause of death is left blank. Given the time of death of 5.40 p.m., it is tempting to speculate that he died of caisson disease, coming up too quickly at the end of his shift, but the evidence for this is missing. Cases where evidence was contradictory or inconclusive were rejected. Checked and double-checked, the figure for the death toll now stands at 73, sixteen more than the 'official' record.

Another mystery was a 43-year-old Italian caisson engineer who died, possibly of typhoid, on 6 September 1885. His name defeated the clerks. Hospital admission records refer to him as Andre or Angels Eggotz, while on his death certificate his name is given as Egatz Arcangelo Di Bortolo. Dr Hunter wrote in his thesis of the case of caisson superintendent Angels Ossi, whom he treated for the symptoms of caisson disease. By 3 September, the day that Angels Eggotz appeared unaccompanied

at the door of the Infirmary Dr Hunter considered that Angels Ossi had recovered. Had he? Whoever he was, he too had a lonely death, his certificate being witnessed by a hospital porter.

There were red herrings. One of the first names identified by the genealogists was that of diver Richard Morton. On checking newspapers and consulting diving historians, however, the team discovered that Morton had been working not on the bridge but on the moorings of the Forth guardship HMS *Lord Warden*. Evidence suggests that previously he may have been involved in searching for the bodies of the victims of the Tay Bridge disaster. The story of Morton has an intriguing sequel. The first patient whom Dr Hunter treated for decompression sickness was the diver sent down the next day to investigate what happened to Morton. On being caught halfway down in a whirlpool of currents, he signalled for help but was brought to the surface too quickly. Dr Hunter commented: 'Some minutes afterwards when I saw him he was gasping for breath, his lips were blue, his face livid and his pulse small, rapid and feeble.'

Three years later the team had the names of over 100 men whose deaths were in some way associated with the construction of the bridge and its immediate aftermath. Now began the process of sifting, confirming and rejecting. The figure as it currently stands is 73 deaths, although the toll may rise as more information emerges.

It is more than likely that the contemporary figure of 57 excluded the men who constructed the railway approaches to the bridge as they were employed by a sub-contractor. Four men died mining the tunnel between Inverkeithing and North Queensferry that carried the line onto the northern approaches of the bridge. The town of Inverkeithing witnessed two funerals within three weeks, those of David Thomson and William Howie, who took over a day to die from severe abdominal injuries. Carter John Millar only had to endure an hour of agony from a fractured skull and ribs after he fell in front of three wagons loaded with stone; miner Thomas Conroy also fractured his skull after falling from scaffolding. A fifth man, 21-year-old labourer Michael Rogers from Donegal, died instantly after being crushed by a landslip during the construction of an embankment on the line linking Dalmeny with the capital.

The official figures also excluded those who died after the Sick and Accident Club was wound up, even if from causes directly attributed to their working on the bridge in the final phase of completion after the official opening. In some cases only modern medical knowledge allows their deaths to be linked to the bridge. Gregarious Fifer and ex-soldier George Fowler died in 1892 from caisson disease, although the death certificate recorded 'supposed heart disease'. He had spent seven years as a Brigger, two of which were as a caisson worker. Although he managed to keep on working until the bridge was completed despite complaining of aches and pains, he was in such poor physical and emotional shape by spring 1890 that he refused to go to the workmen's dinner after the official opening. His son left a note, later discovered by his family explaining why: 'He was very tired of that bridge and didn't care if he ever saw it again. He had had enough.' George then found work as a railway lorryman carting goods from the station to their final destination. He left a widow who gave birth to a stillborn child shortly after George's death. His descendants, now living in

George Fowler, third from left in front row

the USA and traced through the research, still call the Forth Bridge 'The George Fowler Memorial Bridge'.

Six men made the fatal plunge from the bridge in the period between the official opening and the end of 1891. Unmarried painter John Whigham, possibly the first victim of painter's colic, had no family to mourn his passing nor a coffin to fill as his body was never found. On 6 August 1890, 26-year-old carpenter and pattern maker John Burnett slipped while building a gangway inside the girders of the southern approach viaduct. He fell about 150 feet, crashing onto the beach at Newhalls. Burnett, who left a wife and three children, was a popular figure in Queensferry's Masonic circles judging by the attendance at his burial in Dalmeny churchyard. A large procession of freemasons from St Margarets, Queensferry; Kirkliston Maitland; Inverkeithing, St Johns; Douglas, Bo'ness; Kirknewton and Ratho lodges filed behind the coffin as well as members of the Flower o' the Forth Lodge of Free Gardeners.

The question has to be asked – how many other caisson workers succumbed to the effects of the bends? Westhofen referred to the deaths of two caisson workers, both of whom showed symptoms of tuberculosis before going offshore. He attributed their deaths to the rigours of the Scottish winter rather than atmospheric pressure. To date their death certificates have not been identified. On 29 November 1884 the *Dunfermline Journal* reported the 'Sudden Death of a Diver'. Alexander Smart (37) was seized with chest pains on his way home from work. After resting for a time on a wall he attempted to make for a friend's house but collapsed in the doorway. The death certificate recorded 'Angina Pectoris'. He may, however, have been suffering from 'the chokes', a sudden, massive blocking of the pulmonary arterial circulation by bubbles of air due to coming up too quickly from a deep dive.

Another reason for the discrepancy between the official figures and the actual fatalities is that the former only counted men directly employed by Tancred, Arrol and Co., that is, the main contractors rather than the myriad of sub-contractors and individuals who worked under their overall management. In July 1888 Edinburgh's *Evening News* ran a headline: 'Supposed Boating Accident at Queensferry'. It reported the deaths of two nightwatchmen, William Fairley and Thomas Roberts, both of Queensferry and married to sisters Janet and Elizabeth Brown. They drowned while securing navigation lamps on the bridge. Their death certificates classified them as fishermen. Thomas Roberts may have been the same individual who, nine months before, had suffered a fractured skull and serious injuries to his hand when a pail fell from the girders and hit him on the head while he was working as a Brigger on Inchgarvie.

The official figure of 57 also excluded the tragic death of three Briggers in a fire as the Little Couston huts belonged to Messrs Waddell, contractors for the railway approach, rather than to Tancred, Arrol. Erected a mile west of Aberdour, the 80-square-foot building was divided into four dormitories with bed spaces for a total of 106 men, as well as two dining rooms and a kitchen with a single cooking plate. The men, who were largely Irish navvies, paid 4d a night, the doors to the sleeping accommodation being kept locked until midnight to ensure that no one slipped in without paying. About 2.30 a.m. on the morning of 8 October 1888 the deputy manager Mr Scott was roused from sleep by the urgent cry of 'Fire! Fire!' Not stopping even to dress he rushed to the eastmost room, whose door was still locked because of the unruly behaviour of a few of the men. He unlocked it, dashed to the nearest door to the outside and opened it. The gale blew back a cloud of smoke and flames barring the exit. A crowd of naked navvies headed for the front door, which was eventually unlocked by the storekeeper and the man in charge of the canteen. By now some men had escaped through the skylights and others used benches as battering rams to knock down the walls of the hut. Staff and bystanders dashed back in to grab bedding for the naked men. In ten minutes the hut was an inferno.

When the worst was over several men required immediate treatment for burns. Three navvies from the eastern dormitory, however, remained unaccounted for until their bodies were seen roasting in the flames. Two of the victims, James McLachlan and Terence Martin, were lying on the floor while John Ward had not even managed to leave his bed. All were single, middle-aged men whose birthplace was officially unknown. The *Dunfermline Journal* assumed that they were Irish although McLachlan had been brought up in Birkenhead and Martin was believed to have connections with Leith. The remains of the men presented a shocking spectacle. Little more than the skeletons were left and the bones had been bleached by the flames. Later that day the remains were gathered together, care being taken to keep those of each man separate. People from Aberdour brought clothing for the 50 men who had lost their belongings in the fire and parcels of clothes also arrived from Edinburgh and Burntisland. No one, however, came forward to claim the bodies. Their charred remains were put into 'rough but decent' coffins and later interred in unmarked graves in Aberdour churchyard.

In general, little is known of the final resting place of all but a handful of Briggers. Two men, Robert Hughes and Thomas Roberts, lie in the old Vennel churchyard in Queensferry and George Hendry was buried in the New Cemetery in Dunfermline. As a young boy in Davidson's Mains, Arthur Smith remembered watching the hearses pulled by black horses come towards the city. Robert Miller, a bridge timekeeper who drowned after falling from the steam launch *Tay Bridge* on his way back from the Bo'ness Regatta in 1889, was buried with full Masonic honours in Bo'ness. The Millers despatched more than one family member to an untimely grave. After Miller's death his widow remarried pilot George McArthur only to lose him and her brother-in-law David in an accident in 1896. Three years later her 12-year-old son Albert was one of four local boys who went wading off the Binks rocks in Queensferry. The strength of the incoming tide deceived them and Albert slipped off a ledge of rock into deep water. Out of his depth, he panicked and dragged down one of his friends with him. It took rescuers half an hour to resuscitate his friend, but time had run out for Albert, whose body was recovered by grappling irons an hour later.

Over twenty deaths were rejected from the researchers' count because only part of the jigsaw puzzle piece could be made to fit. It is possible that railway labourer Patrick Campbell, who died of gangrene after a double amputation following a fall from a train, was a casualty of the bridge. So little is known of the death of 'Forth Bridge labourer' Patrick McGinty, other than that it was sudden and that a fellow lodger at Smith's Land, Queensferry signed the death certificate with an 'X', that he has been excluded from the count. There is a newspaper report of the accident that befell Robert Anderson, riveter of 2 Big Jock's Close in Edinburgh's Canongate, when he fell from the top of a 20-foot ladder at the Forth Bridge Works in May 1885. When he regained consciousness at least a day later, he was taken to Edinburgh Royal

Infirmary, delirious and severely concussed. Although it was reported 'he is not expected to recover', the search for his death certificate drew a blank. Did he make an unexpected recovery or was there an administrative slip-up in recording his death? Did the stress of working on the bridge result in rope maker George Reid going on a bender in Edinburgh's Old Town where he died from a head wound after falling down steps in the West Bow? He was identified by a letter that he was carrying from his sister in Dundee addressed to him c/o Mr Drummond of Bell Street, Queensferry.

During the period of construction, 12 Briggers were acknowledged to have died from natural causes ranging from a stroke to acute meningitis. The Sick and Accident Club's reports suggest that the number of deaths from natural causes was even higher, reaching a peak of 24 in 1887–8 compared with only six the previous year. The growing workforce must have accounted for the increase, possibly spreading infectious diseases from man to man or to their families. Sixty-year-old joiner William Forrest collapsed with a heart attack on the Forth Bridge Works jetty. Another Brigger in his sixties, storekeeper Robert Wilson, succumbed to a stroke on the vessel taking him to his lodgings on Inchgarvie. While working on the island one young chipper suddenly dropped his hammer and fell to the ground, dying ten minutes later; his death certificate recorded heart disease which Dr Hunter had previously been treating. Heart disease was also behind the death of 45-year-old John Marsden who collapsed in a railway carriage on his way home to Dunfermline after his shift. The question could be asked whether the stress of working on the bridge accounted for or hastened the onset of these fatalities.

One death which could be associated with a Brigger but not with the bridge was that of an unnamed labourer from Kinghorn. He had travelled by train from North Queensferry, possibly at the end of the afternoon shift, and was standing on the platform at Inverkeithing station chatting to a friend when the latter accidentally pushed him against the train as it was starting to move off. He was dragged down between the footboard and the platform. The guard immediately halted the train only to discover his 'horribly mutilated' body. The victim left a widow and eight children.

Another death could be attributable to the bridge but not to the Briggers. On 15 January 1890 a young man, a 'stranger to the district' about 5 feet 4 inches tall and wearing navvy's clothes, was found unconscious at the bottom of a 20-foot railway cutting near Inverkeithing. He died the next day from a fractured skull. Was he a worker or was he looking for work? Who, if anyone, came forward to claim his body? There was even suspicion of foul play. The body of 22-year-old Forth Bridge labourer John Conway was found washed up at the head of Inverkeithing harbour. Marks on his face and elsewhere suggested that he had suffered from more than a fall although a post mortem confirmed that the immediate cause of death was drowning. Witnesses claimed that he had been seen quarrelling earlier but although the sheriff interviewed four men they were not charged with his death.

Was there a cover-up? Although it was financially in the contractor's interest to play down any association between accidents and working conditions on the bridge, there is little evidence to support the conspiracy theorists. It was in Arrol's interest to look after his men as any compensation paid for injury or death was deducted from the contract's terminal bonus. There is a discrepancy of only one Brigger between the official figure of 57 and the deaths identified by the researchers of men working on

the bridge itself during the main construction period. However, without the approach railways the bridge would have been a bridge to nowhere. It also seems unfair to the Briggers' memory to exclude those who died working on vital completion tasks after the official opening. There are inconsistencies even in contemporary figures. At one point Westhofen talked of an average of nine deaths a year over the seven years of the project, suggesting a higher casualty rate than the 57 that he quoted elsewhere. Clerks were often far from perfect. There were accidents that the local police force appears not to have recorded as well as understandable confusions of identity.

The death certificates tell a stark story. The immediate cause of death recorded does not necessarily depict the full sequence of events. On 31 July and 6 August 1888 young bridge labourers John Curran and William Brown respectively fell from the bridge while working on the piers and drowned. No other details of their deaths are known, by then some fatalities not being sufficiently newsworthy to be reported in the press. What occasioned them to fall, from loose planking to a blow on the head by a tool dropped from above, will never be known, nor their final thoughts as they plummeted downwards past the stagings and girders.

Many workers could not swim, reducing even further their likelihood of survival. As early as June 1883, in a letter to *The Scotsman* recommending learning to swim as a way of reducing bathing and boating fatalities, the writer referred to a tragic accident 'only the other day' when a Brigger fell into the water on the Fife side of the river. He thrashed about in the water until another worker who could swim dived in to try to help him. 'In his agony of fear the poor wretch clutched hold of his would-be saviour, and a desperate struggle for life took place.' It failed and both men drowned. As yet a search of the death records has not identified their names.

The death certificates reveal tantalising details of the lives of the men. Just over half were married; sometimes families of eight or more were left bereft of a father and breadwinner. Most Briggers such as riggers, riveters and enginemen worked in labouring occupations but the death toll extended to a storekeeper, four seafarers, a carter and a works gatekeeper who was formerly the colour sergeant of the 2nd Regiment of Foot. The towns of Queensferry, Inverkeithing and Dunfermline each lost ten men and Edinburgh and Leith five men each, although 'usual place of residence' may have been no more than for the duration of the building of the bridge. Although many Briggers came from families of labourers or farm servants, some of their fathers had worked in skilled occupations such as a china painter, a shoe maker, a slipper maker, a blacksmith and a tea merchant. Tellingly, nearly a third of the Briggers' mothers had predeceased their sons, suggesting the high toll of childbirth and poverty.

Thirteen-year-old rivet catcher David Clark was the bridge's youngest victim. About midday on Thursday 13 September 1888, the lad, who lived with his grandmother in Dunfermline, was working on the viaduct near Inchgarvie. He missed his footing and fell over 150 feet, his body bouncing off the ironwork. David died instantly from a fractured spine and skull. His corpse was recovered and laid to rest in the Inchgarvie mortuary until a coffin could be found to return him to his grandmother. At exactly the same time the Duke of Cambridge, visiting Scotland on troop inspection duties, was admiring the bridge from the island of Inchkeith, no doubt unaware of the unfolding drama. Another possible claimant for the title of the

How the men died

Fell	38
Crushed	9
Drowned	9
Struck by falling object	8
Burned	3
Unknown	5
Caisson disease	1

bridge's youngest victim was Thomas Shannon, whose age was variously reported as 13, 14 or 15. The young rivet catcher was probably recruited by his father Patrick: rivet teams sometimes included two or even three generations of the same family. In the late afternoon of 9 December 1888 Patrick was working on a girder on Inchgarvie when something attracted his attention. It was the body of his son who had fallen 140 feet to land almost at his feet. Thomas had been standing on the inner viaduct of the Inchgarvie cantilever when he lost his balance and toppled over the side rail.

It was not uncommon for men to work well into their sixties. One of the oldest Briggers on record was 67-year-old Thomas Sinclair, who fell 50 feet from the scaffolding. He broke his left leg and dislocated his right knee joint as well as suffering severe internal injuries. His accident is likely to have ended his working life. The oldest fatality was 61-year-old road labourer Peter McLucas of Plewlands House in Queensferry, who fell from a cantilever in the early hours of 3 December 1887. Every bone in his body shattered. Too ill to be moved to hospital, he took five days to die at home. His widow Janet was forced to take in a lodger to feed her eight children. Peter left his own memorial, a box made out of wood that he had gathered while working on the bridge. His great-grandson, who still lives in the town, treasures this memento of a lost life.

Records also help to piece together the lives of individual Briggers. The Duffy family first emerge in Queensferry in 1851 thanks to the careful recording of their names, ages and occupations by the census enumerator. The duties of this enumerator were time-consuming. He first had to find the cramped and insanitary home of Thomas Marin in the warren of lanes and back courts on the landward side of the High Street. He then had to account for each of the eight adults and six children, all Irish-born, who made up the household on census night. Among them were 41-year-old widower James Duffy, who worked as an agricultural labourer, and his two sons Bernard and Patrick, aged twelve and ten. It is probable that James, like so many others, had been driven to escape the poverty, disease and starvation that his wife and other children may have succumbed to during the potato famine. Even Thomas Marin's overcrowded rooms were a better option than the devastation that he left behind in Ireland.

By the next census clerk's visit in 1861, life for James Duffy was looking up. He had a home of his own and two children living with him. His eldest son Bernard lived next door with his wife Rose. Ten years on and Rose had changed her name to Carrah, later also described as Kerracher or Carrechar, although there is no record of a husband. She had three children living with her, Hatty and Patrick Duffy, and two-year-old Michael Burrigan. Whether Michael was born out of wedlock, if indeed she was married, or the child of a friend or neighbour that she was labouring up will probably never be known. His father was labourer Richard Berrigan. By 1881 Rose was a widow working in the fields and living with her sons. James was working in the local shale mines while Patrick was a baker's boy and Michael unemployed. Sometime in the next few years Michael found a job as a labourer, his recorded occupation when, on 16 December 1883, he was caught poaching on the Dundas estate with five teenage friends. He had the option of a fine of 6d plus 7s 6d expenses or of seven days in prison. It is not known which he chose. The cooking pot must have been empty that Christmas.

Sometime before 9 May 1887 Michael started work as a rivet heater on the bridge. Sadly the story ends here as by nightfall he was just another statistic. While working inside a vertical column on Inchgarvie that morning he was climbing a ladder when a plank became dislodged. He fell headlong 20 feet and died almost instantly from a broken neck. Two days later James Duffy, who described himself as Michael's brother,

registered the death of Michael 'Berrigan' or 'Bearrigan'. It may have been the same James Duffy who was later charged with breach of the peace and assault, incurring a fine of 14 shillings or 14 days in prison. Were it not for the research carried out to resolve the confusion over the names, the story of Michael's short life might never have been told.

Although Michael's death did not merit even a paragraph in the local newspapers, the press had long since discovered that accidents made good copy: 'Fatal Accident at the Forth Bridge', 'Another Fatal Accident at the Forth Bridge'. Newspaper editors helped fuel public concern about the safety of the Briggers. Three accidents less than two months apart in the summer of 1887 and each involving two fatalities were critical to making the bridge a cause célèbre. In early June riveters George Hendry and Martin Welsh died and four men were injured when a poorly secured section of scaffolding collapsed. As they fell Hendry and Welsh struck a girder which fractured their skulls; they were dead by the time they hit the water. Two men were rescued from the sea and the remaining two managed to grab onto the struts until help arrived. Most of the injured were in such a bad way that a telegram was immediately dispatched to the Infirmary in Edinburgh advising staff to make beds ready. Jess Winters was too ill even to move and remained in the Queensferry ambulance house under the care of a doctor while young William Burgess was deemed well enough to travel by train to have his wounds dressed.

At the end of July a plater and his team of four helpers were fitting a cover plate 300 feet above the river. A rope, which was too loose to secure the plate adequately, snapped under its weight and the plate headed downwards slicing through the planking of the staging where two drillers, John McEwan, or Gallacher, and John McClinchie, were working. Only minutes earlier a visiting party of railway managers had passed close by. On spotting the two Briggers one visitor remarked that he 'would not work up there if you gave him £1,000 a minute'. They left minutes before the accident.

Within a few days yet another two Briggers were dead. Six men were painting 130 feet above the Forth when, about midday, their platform gave way. Three men succeeded in clinging onto the girders above them and were soon rescued by fellow workmen who tore up the flooring above the staging to provide a way to safety. Two Briggers, James Syme and teenager John O'Neil, crashed onto the jetty, breaking their necks. A third man, 40-year-old William Watson, fell feet first into the water and suffered no more than severe shock. A plank hurtled through the roof of a shed on the jetty and landed beside the blacksmith, who was reported to be unharmed by his experience.

In autumn 1887 the *Dunfermline Journal* ran a campaign against what it described as the 'slaughter' on the bridge. Letters to *The Scotsman* and the *Edinburgh Evening News* called for harnesses and safety wires to be used. In retaliation a representative of Tancred, Arrol and Co. accused the press of 'gross exaggeration' in reporting the number of deaths as exceeding a hundred, claiming that up to September 1887 only 20 men had lost their lives. Half of the deaths were routine accidents that might occur in any workshop or yard. Tancred, Arrol's representative cited the example of joiner Edward Davies whom the press had reported as being killed by a falling metal bolt. 'The fact is the case was one of a scalp wound, but not serious.' The writer concluded: 'All that appliances, safeguards and supervision can

do, we try our best to maintain, but we cannot successfully contend against recklessness or thoughtlessness of the men themselves.' He appealed to the press to avoid giving 'a sensational character to many insignificant mishaps' as this had resulted in anxious letters from relatives 'from the colonies and other distant parts' months after the event. Three days later, Davies died in hospital of a compound fracture of the skull.

The committee of the Sick and Accident Club sought to put the record straight by publishing their statistics in *The Scotsman*. 'Greatly exaggerated statements in regard to the number of accidents – fatal and otherwise – having appeared in the public press recently, the committee subjoin a record of all accidents which have occurred since the works were commenced in 1883.'

	Fatal	Infirmary	Minor Cases	Total
To Sept 1884	2	6	30	38
1885	6	20	86	112
1886	4	19	109	132
1887	17	22	132	171
Totals	29	67	357	453

In a speech delivered in Dundee at the beginning of November, Benjamin Baker pointed out that the accident rate was insignificant compared with coal mining. The death rate on the bridge was 1 in 300 of the workforce compared with goods train guards and brakesmen where the rate was 1 in 192. A modern comparison also helps to put the death toll in context. In proportion to the size of the workforce, the death rate for the Forth Rail Bridge was no higher than for the Forth Road Bridge, constructed in the early 1960s.

In response to the furore the government commissioned Major-General Charles Hutchinson, responsible for quarterly inspections of the bridge, to suggest safety improvements. The editor of the *Dunfermline Journal* was not prepared to lay the matter to rest especially as another six men had been killed during the autumn of 1887: 'The monthly slaughter cannot be tolerated – the warnings by the Government Inspectors and the contractors seem to be allowed to be blown over the cantilevers like the morning mist.' He proposed that the quarterly inspections should not confine themselves to the quality of the workmanship but should address health and safety concerns through a thorough investigation of any accidents as in the coal mines. He advocated the appointment of an independent inspector, as existing accident reports to the procurator fiscal only addressed the how and not the why. Benjamin Baker was moved to write a dusty reply pointing out that four Royal Commissions dealing with accidents had rejected this solution as appearing to interfere with management or indirectly to guarantee machinery and safety arrangements that might be faulty or inadequate.

The Scotsman claimed first-hand evidence of carelessness, taking the view that at least one of the recent fatal accidents could have been averted by stricter supervision of the workforce at the main piers. 'We ourselves noticed when on the top of the North Queensferry pier, an accident due entirely to the carelessness or ignorance of a foreman. This providentially had no ill results, but it might easily have caused loss of life. No expense should be spared in order to provide a sufficiently numerous and experienced staff of overseers.'

One London journalist from the influential *Pall Mall Gazette* visited 'the worker's village of South Queensferry' in April 1889 to investigate the situation for himself. He obtained an interview with William Arrol: '"Accidents!" said Mr Arrol, "there are not more accidents here than in an ordinary shipbuilder's yard . . . But there's only one Forth Bridge, and as everything that takes place seems to get reported in every paper in the country, people get quite erroneous ideas about the fatalities that occur."'

The journalist then took a lift to the top of the bridge, 'swinging like a caged canary at the end of a thin rope, higher and higher and higher in mid air.' After admiring the panoramic view, he called in on Dr Hunter 'whose house and surgery at the other end of South Queensferry had to be reached by traversing the main, if not the only, street of that queer, half-old, half-new, little town'.

He quoted Dr Hunter as estimating, 'Including five drowning cases, the fatal accidents from all causes in connection with the bridge amount to 53; but if you take simply the number of those killed in the actual construction of the bridge, there have been 48 lives lost, death taking place either at the time of the accident or soon after.'

The reporter concluded: 'So much, then, for Dr Hunter's statistics, which I have quoted, dry as they may be, for the sake of showing what is the nature of the "casualties" that occur in a great engineering work like this, and what is the price to be paid, in blood and human life, for a big bridge, even when all is done to drive as hard a bargain as possible with the death angel.'

A final salvo was fired in 1890 by no less a person than the Prince of Wales, in response to a letter that he had received from a correspondent in Dundee on the day before he officially opened the bridge. The letter drew his attention to the plight of the widows and children of the Briggers. He instituted an investigation and published its findings in *The Scotsman*. The letter stated that there were 56 deaths over the seven years of construction, a figure probably taken from the report of the Sick and Accident Club. It claimed that ten of the victims had left no near relatives to whom to offer compensation. It vindicated Tancred, Arrol's safety record: 'It appears that exceptional precautions were taken to guard against every form of accident.' The letter described the platforms as 'substantial, well fenced-in, and secured', the access as being 'as convenient as possible and well provided with handrails' and the purpose-designed lifts as being fitted with safety appliances which never had to be called into use. It went on to consider the precautions taken against the risk of falling tools and material including the provision of receptacles which the men could use to store loose items like rivets and safety wire-netting screens to catch tools as they fell. The letter concluded: 'His Highness is assured that owing to the precautions adopted the number of casualties which terminated fatally is remarkably small.' All that remained for the prince was to send his sincere condolences to the widows and their families.

There the matter rested, the issue of the human cost of the bridge being rapidly buried under the paeans of praise for its engineering. Public memories were short while the private memories of the Briggers' families were long.

CHAPTER 5

RESIDENTS, ROYALS AND REVELLERS

Writer and traveller Francis Groome may have been ambitious in his estimated timescale for the completion of the bridge but by 1885 the impact of the works and the hundreds of offshore commuters was being felt in the communities on both sides of the Forth. It was not only the physical landscape that was changing – the first brick houses on the hill overlooking Queensferry, the new jetty at North Queensferry, the ever-changing views from windows which would never look out on the open estuary again – bridge operations placed new pressures on basic resources like water, which were already severely over-stretched. The workforce provided new opportunities for local businesses, especially innkeepers and publicans. The narrow streets were unaccustomedly crowded with men coming off shift or tourists pouring off the steamers to marvel at the giant piers with their ever longer arms stretching towards the final handshakes. Despite it all the rhythm of daily life in both communities continued as before with council and parish meetings, sports days and dances, domestic dramas and kerbside gossip.

Most residents on both sides of the Forth greeted the news of the building of the bridge with enthusiasm. Shopkeepers extended and renovated their premises and new businesses opened in anticipation of the invasion of workmen and visitors. Queensferry smartened itself up with the removal of old walls and derelict buildings. The downside, however, emerged as early as 1883 when contractors started to remove sand from the foreshore to make concrete. After complaints about damage to sea walls and dumping of rubbish, they had to transport sand from banks exposed at low tide 10 miles downstream off the Fife coast. Housewives must have increasingly cursed the bridge, their ears assaulted by the constant din of riveting and their washing regularly covered in smuts and later spots of paint.

One of the most immediate effects of the bridge came at dusk. Residents would have watched for the lights coming on at both ends of the bridge and on Inchgarvie. Electricity was not a domestic luxury that they would enjoy for many years to come. Views of the bridge at night varied over time as, thanks to his own ingenuity and the unique Siemens system, the resident electrician was able to add or move lights in line with the progress of the work. Philip Phillips described the vista: 'The effect of this

It remains to be seen what will be the effects of the completion, about 1888, of the great Forth Bridge.

Francis Groome, 1885

illumination stretching from one side of the river to the opposite shore thus showing the outline of the bridge on the darkest nights, was a sight both grand and unique, and viewed from some miles either up or down the channel.' The new lighthouse on Inchgarvie added its beam to the night sky, joining those which had guided vessels in the Forth for many years.

Indeed both Queensferries owed their existence to the vessels that had regularly plied across the estuary since 1130. By the 1880s North Queensferry was 'a favourite summer resort for sea-bathing', its public facilities running to little more than a post office, coastguard station, Free Church and public school. In the previous half century its resident population had fallen by nearly a fifth to 360. It was overshadowed by its larger industrial neighbours Inverkeithing and Dunfermline, where many Briggers lived or lodged. Smaller communities on both sides of the Forth also saw an influx of temporary residents; one theory as to why the village of Kirkliston acquired the nickname 'Cheesetown' is because of the number of packed lunches or 'pieces' prepared for the Briggers as well as for local shale miners and agricultural workers.

The community which experienced the greatest impact during the building of the bridge was the royal burgh of Queensferry. In 1881 it had over 1,000 residents with almost as many again living in the surrounding agricultural and shale-mining villages, and on the country estates that hemmed in the town on its three landward sides. In the peak years of construction, thousands of Briggers more than doubled the number of people on its narrow streets, albeit Queensferry had prior experience of dealing with a transient population as nearly 300 sailors were based on the guardship moored offshore. An armed naval vessel had patrolled the Forth since the mid 1850s. It also acted as a training base for reserves and for the East of Scotland coastguard service. In 1886 the elderly HMS *Lord Warden*, one of the heaviest wooden ships afloat thanks to her steam engine, machinery and iron cladding, and with a notorious reputation as a 'roller', was pensioned off. She was replaced by HMS *Devastation*, which belonged to the first class of Royal Navy battleships to dispense entirely with sails. As a mast was no longer necessary, gun turrets could be mounted with an unobscured line of fire.

The royal burgh had once been a wealthy port, boasting 20 ship owners in the mid seventeenth century. By the 1880s they had long since gone, although the harbour still handled a coastal trade in coal, manure, barley, stone and potatoes. Whisky distilling and trawling were the main employers but stocks of herring were dwindling due to over-fishing. According to the author of *Summer Life on Land and Water at South Queensferry*, the town also had aspirations as a seaside resort. 'Ladies and gentlemen, as well as families of children, distinctly understand where they ought respectively to repair for bathing – females and children being permitted the full range of the beach, along the town, and for some distance eastwards and westwards beyond the pier, whilst the gentlemen willingly move off to a greater distance so that no misunderstanding can occur.' The economy of the town had been in the doldrums for some time and so great hopes were placed on the bridge to revive its fortunes. These hopes started to be realised once construction began: 'At present all houses are occupied, and there is a stir in the place by the introduction of so many workmen into it, which is something new in its history.'

The town still retained its medieval layout, its narrow main street following the line of the foreshore for half a mile. Short alleys led off it to the harbour and to warrens of houses and vegetable patches. The Irish quarter was notorious for its insanitary slums. Town facilities included the seventeenth-century parish church with its damp and over-crowded graveyard and a United Presbyterian church, a post office, a branch of the Clydesdale Bank, four hotels and a townhouse where the town council met to administer local affairs. The elected council, consisting of the provost, two bailies and six councillors, oversaw the work of burgh officer Thomas Hunter, who earned a salary of £8 a year. He also acted as the officer to the commissioner of police, his remuneration being made up of £1 per annum plus a fourth part of the fines recovered by him as prosecutor under the legislation governing public houses, an arrangement that potentially was a recipe for corruption.

The town council was backward-looking, dilatory and determined at all costs to keep local taxes low. The burgh had already declared itself bankrupt in 1881 but dug itself out of the hole by increasing rates. There were regular complaints about the insanitary state of the streets, not helped by the unwillingness of the refuse collector to include Catherine Bank, which housed Briggers and their families, on his rounds. The issue of an adequate supply of water highlighted their lack of investment in even basic facilities. Traditionally the town had relied on water collected in local fields being brought in to augment the private wells, which regularly had to be closed on suspicion of being tainted with sewage. Improvements in domestic sanitation resulted in increased consumption without any balancing increase in supply.

Then the Briggers and their thirsty steam-driven machines arrived. On the Fife side scarcity of mains water soon forced the building of a storage tank at North Queensferry. Water for Inchgarvie came from the mains on either side for the Forth.

Overleaf. HMS *Devastation*, Forth guard ship

On the south side, water for machinery was taken from the pits of the Dalmeny Shale Works, first passing through gravel beds to remove the grit. Drinking water came from a local seam of sandstone but the supply was often intermittent or failed entirely. Housing for bridge workers added to the pressure. As early as June 1884 the Queensferry police commissioners agreed to supply water to the new Catherine Bank houses only if there was a supply available. Tancred, Arrol and Co. had to contribute 10s for each of the sixteen taps while the individual households had to pay 1s a year for drinking water.

The situation came to a head in the summer of 1885 when there was a 'water famine', a situation repeated over the next two years. A drought resulted in water failing to flow from taps in streets and houses because the level in the reservoir had fallen below that of the delivery pipe. The council stationed a watchman to guard the remaining water in the reservoir, which amounted to less than a week's supply, and to fill up containers which people then carried into town. Tancred, Arrol came to the rescue by erecting storage tanks at the harbour which were filled with water transported by steam barges from the Starley Burn 7 miles downstream, near Aberdour. Everyone could use the water free of charge. Residents clubbed together to open a well in the doctor's garden. The council breathed a sigh of relief that there was temporarily no guardship in the Forth as they received £80 a year from the Admiralty in return for supplying the vessel with water.

The drought of 1887 was particularly severe, with people resorting to washing in sea water. The pump well had to be reactivated and the water boats resumed their run to Aberdour. A new, permanent, water supply was urgently needed. At Dr Hunter's suggestion, feelers had been put out to the neighbouring communities of Davidson's Mains, Cramond and Kirkliston to invest in a joint scheme. In late 1887 the parishes applied to the Secretary of State to acquire the necessary land to pipe in water from the Pentland Hills. Despite accusations that a petition in support of the scheme included the signatures of non-resident Briggers, a local inquiry found in favour of the scheme. Last-ditch objections by local landowners were overruled and the town finally won the right to a reliable supply of clean drinking water.

The police authorities were rather more forward-looking in preparing for the Briggers, no doubt anticipating hordes of drunken navvies rampaging through the town. In the late 1870s the police station in Wester Park had been doubled in size 'in the expectation of the Forth Bridge causing a large addition to the inhabitants and so accommodating demand for admission'. The jurisdiction of the magistrates was extended in 1882 to cover the area around the bridge. Police manning was also an issue. Following complaints about the conduct of Sergeant McLeod, especially 'the difficulty of other officials in getting on with him', the chief constable replaced him with 'an active and experienced officer' with the rank of inspector to be supported by an additional constable. The chief constable, however, also took the opportunity to complain about the state of lawlessness in the town. The adequacy of arrangements for policing remained on the agenda after work on the bridge started. In 1885 the West Lothian police committee reported that 'owing to the great influx of people in connection with the Forth Bridge Works the necessity existed for increased police accommodation'. It was decided to build a temporary police station close to the works, with three cells and living quarters for both a married and a single constable; Tancred, Arrol agreed to fund half the cost. An extra constable was also drafted into

North Queensferry to cope with the increase in troublemakers.

Queensferry already had an 'unenviable reputation of being the most disorderly' of West Lothian's five police districts. In 1878 there were 223 convictions for crime in the town. The extent to which the Briggers were responsible for any increase in crime remains a matter for debate as police court records often describe the accused simply as 'labourer'. William Fox and John Drummond, who were sentenced to nine months in jail for the 'criminal assault' of two young Queensferry women, could have been bridge, farm or shale-works labourers. Was Farquhar Swan from Greenock who stole a suit from a house in North Queensferry in 1885 looking for work as a Brigger?

Some Briggers, who can be positively identified as such by their occupation, found themselves on the wrong side of the law. Teenage rivet heater Henry Wilson turned highway robber when he stole a silver watch from a man on the road near Newhalls. When PC Dixon tried to prevent two drunk Briggers from boarding the workmen's boat at North Queensferry, he was rewarded with a blow to the face. The pair were later given the choice of 30 days in jail or a £5 fine and a caution for good behaviour over a six-month period. At the end of construction, Jeremiah O'Driscoll was charged with stealing tools and resetting them. Briggers also committed crimes against fellow bridge workers. Railway navvy Frederick Elliot found himself in jail after stealing from the workmen's huts. A holder-on and a labourer set upon a fellow worker in North Queensferry, battering his head with a stone. The victim, engineer John Grey, bore 'marks of severe handling' and ended up in hospital in Dunfermline.

Forth Bridge workers were also the victims of crime. Contractor Don McLaren had a rather lean Christmas after labourer Thomas Powell stole his roast turkey. Was the man found lying drunk in the road east of Queensferry a Brigger and what about the thief who stole his boots and tried to sell them? There were also instances of white-collar crime. The *West Lothian Courier* reported a racket popular with 'the varied specimens of the *genus homo*, known familiarly and poetically as the "black squad"'. The case in question involved the attempt by a group of navvies to raise funds for more drink by asking a foreman for money to bail out a friend from the police cells. The foreman was not hoodwinked and wrote a note to Inspector Mackay offering personally to guarantee the good behaviour of the prisoner, if released. A more serious case of fraud resulted in Inchgarvie foreman W. Torrance being sentenced to a minimum of two months in jail. For six consecutive months in 1888 he had overstated the pay due to some of his workmen in his paperwork for Tancred-Arrol, later retrieving the excess pay from the men, probably in return for a backhander.

Many crimes – breaches of the peace, assaults and affrays – were probably fuelled by drink. Not all could be blamed on the Briggers. While first-footing early in the hours of New Year's Day 1884, local fruiterer John McArthur was so incensed by elderly Mrs Bowie complaining of the noise he and his friends were making that he hit her on the face and twisted one of her fingers until it broke. On the night of 3 August 1889 a riot took place in the town after the police arrested Irishman James O'Donnell 'who was behaving in an outrageous manner'. His fellow countrymen went to his defence, resulting in a pitched battle that left one police constable unconscious. Briggers may have been in the thick of the battle.

Queensferry was well endowed with places to buy drink, having catered for travellers and carriers waiting for the ferry for centuries. By 1878 the immediate

Overleaf.
Travellers approaching the ferry at Hawes Pier

The infamous Hawes Inn

locality boasted six hotels and inns, eight public houses and 14 licensed grocers. Because of its location next to the works the Hawes Inn attracted most opprobrium. Benjamin Baker remarked: 'It flourishes too well . . . its attractions prove irresistible for a large proportion of our 3,000 workmen . . . The accident ward adjoins the pretty gardens with hawthorns, and many of the dead and injured men have been carried there who would have escaped had it not been for the whisky of the Hawes Inn.' At least one man, a mason who was probably a Brigger, died as the direct result of drink. A few days before Christmas 1889 the driver of the 8 p.m. train from Edinburgh to Queensferry reported that his engine 'had passed over some object on the line, near the Forth Bridge workmen's platform'. A search party soon discovered the fearfully mutilated body of Frederick Duff, the train's engine having literally cut him in pieces. Earlier that evening he had been stopped from wandering about on the line in a state of intoxication. He had then fallen asleep but on awakening had returned to the tracks and his death.

Initially at least, the Briggers did not need to seek out the local watering holes as drink was on sale in the three works canteens on either side of the river and on Inchgarvie, even on Sundays. As early as September 1883, the authorities had to issue a handbill warning the men that anyone found drunk in the Queensferry canteen or conducting themselves in a disorderly manner would be barred for a month. A second offence would result in instant dismissal. The warning proved ineffective and a month later the sale of drink in the canteen was temporarily banned on Sundays. The *Dunfermline Journal* voted the experiment a success: 'Formerly, the incapables

who loafed around the canteen were a disgrace to themselves and to all who permitted such a state of things.' Gradually the rules were relaxed, the men being allowed at first to have only one pint of ale and only with a meal. Thereafter the bar was open for three hours a day with no restrictions on the quantity drunk until the Justices at the Linlithgow Quarter Sessions threatened that the licence would be withdrawn entirely if the canteen did not revert to the practice of one pint with a meal.

The canteen at North Queensferry came under similar criticism during the trial of riveter James Trainer for hitting the boatman at the end of the pier. Trainer, who lived in Queensferry, had crossed the Forth to escape the restrictions on his local canteen. He pled guilty, confessing that he had purchased half a gallon of beer and some rum and was too drunk to recall the incident. The procurator fiscal pronounced that 'North Queensferry was bad enough on Sundays with the drunken bridge workers who were resident in the village without the contingents that sometimes arrived from the south side.'

There were complaints about drunken rowdiness on the south side of the river too. Captain Kennedy of HMS *Lord Warden* wrote indignantly to *The Scotsman* about the deplorable state of affairs, which he attributed largely to the influx of the Briggers. 'This quiet and once popular little watering-place is fast being turned into a rowdy and disreputable hovel. The streets and suburbs are now infested with drunken vagabonds who insult the peaceful inhabitants to such an extent that it is no longer safe for ladies to walk alone in this neighbourhood . . . There are no less than twenty-three licensed houses for the sale of intoxicating liquor. These establishments are doing a roaring trade, especially on Sundays, on which day the number of drunken loafers is most conspicuous.' He called for the three-man-strong police force to be increased and the removal of licences from publicans who sold drink on Sundays, otherwise he would be forced to 'land a picquet from the guardship for the protection of ladies and the support of the police'.

After the Queensferry magistrates revoked the licences of local hotels to sell drink on Sundays the exodus north was not confined to the works canteen. Sheriff Gillespie forecast: 'Rowdies will come from the other side to North Queensferry to get drink, and the place will be made quite intolerable for quiet people living there.' He was not entirely wrong in his assumption. On Sunday 2 September 1888, for example, William Sawers assaulted a landlord who was trying to eject him from his premises. The authorities in Dunfermline were also concerned, blaming the Briggers for the rowdiness in the town especially on Saturday nights after the pubs closed. One Monday in 1886 a group of Briggers started to brawl in the High Street. When Constable Hunter tried to intervene six men brought him to the ground and kicked him. Hunter managed to summon assistance by blowing on his whistle while trying to ward off blows with his baton. Trouble even spread to the capital where policemen used to patrol the notorious Grassmarket area in groups of six for protection.

The railway navvies had an even worse reputation than the Briggers. When they arrived to construct the approach railways, trouble escalated. Sunday shebeens in the Dalmeny railway huts were an additional port of call both for drinkers and the police, who on one occasion confiscated an 18-gallon cask and nine dozen bottles of beer. A similar shebeen operated in the huts at Jamestown, outside Inverkeithing, where, despite being attacked with a poker and having stones hurled at them, the police managed to remove six casks of beer and ale. A riot broke out in April 1889 when the town's senior bailie waved a truncheon and other town officials had to come to the aid of the police. The *Queensferry Observer* noted: 'When the town was frequented by large numbers of the "black squad" nothing like Saturday night's disturbance ever occurred,' but resigned itself to trouble on the streets until the southern approach railway was built.

The arrival of Galloway's excursion steamers at Queensferry in the summer of 1886 added to the problem. The erection of a 300-foot wooden extension to the existing pier allowed cruises from Edinburgh to call in at the town as well as providing residents with a local stop on the Stirling-to-Leith shuttle service. Weekend cruises from Leith and Aberdour on the paddle steamers *Lord Morton* and *Edinburgh Castle* promised views of the bridge, the works and the guardship HMS *Devastation* for only sixpence a trip. There were also Tuesday and Friday evening cruises to the Forth Bridge Works. Until a barrage of complaints forced the Forth Bridge Railway Company to reduce the price to a penny, visitors had also to pay sixpence to land at the pier unless travelling on one of the company's boats. At the town council's insistence Galloway had to provide a urinal on the pier.

Sunday excursions on Galloway's *Fiery Cross* proved particularly popular. The main attraction for Edinburgh drinkers was not the bridge but the hotels as the law disallowed the sale of alcohol on Sundays except to bona fide travellers. A sail on the Forth was long enough to convey traveller status to the thirsty who followed it by a largely liquid supper. Queensferry's magistrates finally acted in 1887 by downgrading

the hotel licences of the Hawes Inn, the Stag Inn, the Forth Bridge and the Queensferry Arms Hotel to that of public houses, thus neatly barring Sunday drinking by the back door. The publicans appealed. Despite letters of support praising the management of the Hawes Inn and blaming day-trippers who brought their drink with them for any trouble, over 80 per cent of householders signed a petition supporting the ban. The publicans lost their appeal and threatened to take the issue to the Court of Session. The ban, which meant that the town no longer had any accommodation for genuine travellers, helped to open up the market for temperance hotels. By the end of the century the Forth View, the Caledonian and Bradley's Temperance Hotel were doing a brisk trade.

Other than drinking in the pub, visiting Mr Kell's shooting galleries and watching the local football team, the Bellstane Birds, entertainment in Queensferry was limited and unlikely to have strong appeal to hard-drinking Briggers. In winter, musical evenings were laid on for the workmen in the large dining hall at the Queensferry works, which also hosted the Industrial Exhibition for the Parishes of Dalmeny, Cramond, Queensferry and Kirkliston in 1888. Its star attractions were busts of John Fowler and William Arrol. There was a Young Men's Mutual Improvement Society and a Musical Association. Less uplifting ways of passing the time resulted in the minister having to chide his flock for gambling in the church on Sunday evenings. It is likely that prostitutes sold their services on the streets, retreating with clients up the back alleys. One bridge engineer from respectable Bridge of Weir near Glasgow attended the Infirmary with gonorrhoea while a Queensferry woman was in and out of the same institution for treatment for syphilis. One break in the monotony was the Ferry Fair held on a Friday in mid August. On the eve of the Fair the 'burry man', covered from head to foot in sticky burrs, paraded the streets. As well as merry-go-rounds, shooting galleries, Aunt Sallies, sweet stalls and a wheel of fortune in the High Street, sports were held in the Bellstane Birds' football field. Entries for one 220-yard handicap were confined to Forth Bridge workmen, watched in 1889 by Sir John Fowler with a party of local aristocrats and dignitaries. Briggers are also likely to have entered the open events.

Addressing the Briggers' spiritual needs was an early priority. Two local Church of Scotland ministers brought religion to the men by holding Sunday services in the Queensferry canteen, although they had to contend with the smell of liquor from the bar at the other end. The same year a group of Roman Catholics started to raise funds to build a chapel for the large influx of Irish workers; Father Turner from Davidson's Mains was already conducting Sunday mass in the canteen. The next year the diocese purchased a corrugated-iron chapel from a Baptist congregation in Leith which could seat over 300 worshippers. It was erected in the Loan near the railway station with an adjacent chapel house for Father Turner.

The day-to-day business of the two communities continued despite the hammering of rivets. Captain Dundas of Inchgarvie offered to convert the remaining fragment of Queensferry's Carmelite friary into a public reading room. Two years later Mr Anderson cashed in on the town's new-found prosperity by building a row of four shops, his wife later being admonished for using the pavement as an additional display area for her goods. Traffic across the Forth was disrupted in July 1887 when an excursion train from Peebles went off the rails near Port Edgar and again six months later when the ferry was grounded by an exceptionally low tide, both timely reminders of the benefits to come from the bridge. The railway station was broken into and the contents of a trunk of boots and clothing strewn along the platform. The 20-foot-long carcase of a bottle-nose whale was discovered on the beach, while a live one was observed blowing off Inchgarvie. The post office in North Queensferry threatened to sell a lost black-and-tan collie if its owner did not claim it in the next five days. Yachtsmen noted that the annual North Queensferry Regatta had added 'and Forth Bridge' to its title.

There was the eternal rhythm of births, marriages and deaths. In 1886 George Reid the coal merchant became the proud father of a girl, appropriately at the house on the High Street known as Black Castle. The bridge itself brought romance when in 1887 Wilhelm Westhofen married Alexandrina Glendinning, daughter of the factor of the Dalmeny estate, a woman almost half his age. A week before Christmas 1884 one of Queensferry's oldest residents, Mrs Dowd or Carrigan from Echline Farm, passed away at the reputed age of 103. Originally from Ireland, she could still read without spectacles and she knitted stockings until a few days before her death. Widow Margaret Brown or McArthur tried to take her own life by jumping off Queensferry pier. As attempted suicide was a chargeable offence, she then faced the ordeal of a trial at Linlithgow Sheriff Court. She pled guilty and was admonished by the sheriff. Why she was driven to take this drastic action was not revealed.

Divorce was guaranteed to set tongues wagging. In November 1883 the talk of the town was not just the new electric lights at the works but the amorous affairs of the local minister who was accused in court by his wife of intemperance and infidelity. On one occasion he had locked himself in his study for an hour with a young woman, later a servant at the manse, who was supposedly helping him update the communion roll. The gossips continued to have a field day when he was forced to resign his charge, protesting his innocence to the end and that 'his accusers had cast him into the world, a beggar in his old age'. The final twist to the saga came when one of his young ladies took him to court to claim support for her illegitimate child.

Drama was not confined to the land. In the summer of 1884 a Danish schooner laden with pig iron from Grangemouth ran aground on Inchgarvie. The Forth Bridge steam tug *Ruby* towed her into Queensferry harbour, where it was discovered that her hold was under 6 feet of water. Two years later the recently arrived guardship HMS *Devastation* broke from her moorings and had to lie up at Hopetoun until a boat could be summoned from the Thames estuary to repair her mooring chains. The same year Queensferry harbourmaster Colin McArthur drowned after his lug-sail boat was caught in a squall. He might have made it safely across before the storm broke had three Briggers not removed the boat from its North Queensferry moorings and beached it. A yachting trip by two lawyers to view progress on the bridge ended in tragedy. The men had been spotted waving at a passing steamer but were assumed to be bathers as they appeared, albeit wrongly with hindsight, not to be in difficulties. On a summer Sunday afternoon in 1888, eight young North Queensferry men rowed up the river to Rosyth Castle where six of them landed. Then the boat capsized in unexplained circumstances and it was only due to the prompt action of the local sheriff and another bystander that the remaining two lads survived. Two brothers were attacked with cramp during an outing to Longcraig Pier with the Forth Bridge Swimming Club in July 1887. Only one survived.

The construction of the railways posed other dangers for the citizens of

Briggers and guests await the arrival
of the Shah on Inchgarvie

Queensferry. The late 1880s also saw a spate of drownings in the old quarries to the west of the town which had been dug out during the building of the line to Port Edgar. The decomposed body of John Sheppard was only identified by the name on the fly leaf of his pocket New Testament. Weeks later, while strolling near the quarry, Patrick Gallacher heard a young woman screaming. He was just in time to rescue a child who had been attracted to the stagnant water by spotting a lady's hat floating on the surface. Later that summer, labourer James Scott from Forfar drowned in the quarry on his way to catch the steamer, his companion vainly having tried to rescue him. After a second fatality moved the commissioners of police to write requesting that the quarry be properly fenced off, the secretary of the railway company retorted that local people were breaking the law by ignoring the 'No Trespassing' notices.

The Forth provided more pleasant sights for both Briggers and locals, such as the visit of the Channel Fleet in October 1887, during which occasion Tancred, Arrol offered free cruises on the work boats.

Apart from the ubiquitous *Scotsman* reporter, only local people and workmen were privileged to witness one momentous event. The passing of the first railway engine over the tracks on the morning of 21 January 1890 had been kept secret from the public. 'It came very much as a surprise to the people of Queensferry to see in the early forenoon two great trains moving abreast at a slow pace up on to the parallel lines of the south approach viaduct . . . Everyone was soon at the door to see the unusual sight, and many a glance was cast in the direction of the bridge during the remainder of the day.' When the trains, pulling mineral freight twice the weight of what the bridge would ever carry in practice, halted for over half an hour on the viaduct, there was much speculation as to what was happening. In fact the delay was due to the track on the north side not being quite ready. Progress continued 'at walking pace' with frequent stops amid showers of hail and eventually the crowd lost interest. The engineers, however, were delighted, comparing the pressure on the bridge to the 'swaying of a tree in a gentle wind'. Three days later the first passenger train crossed with the directors and senior management on board and the Marchioness of Tweeddale at the wheel, probably making her the first woman train driver.

Visits from royalty, eminent engineers, and foreign visitors became a regular feature of life in the two communities. The first royals to inspect the bridge were the Prince and Princess of Wales on 24 August 1884. At the Forth Bridge Works three foremen presented the princess with a paperweight made out of whinstone, granite and steel to symbolise the materials of the bridge. The royal party then elected to walk down the steep incline to the jetties rather than ride round by the road. Two hundred Briggers stood on top of one of the caissons and cheered as the royal party crossed the Forth to the boom of a gun salute from HMS *Lord Warden*, festooned with bunting for the occasion. The ground in North Queensferry was deemed to be too

The Shah's visit. Fowler is second from left

rough for the ladies to disembark but the prince enthusiastically climbed to the top of one of the piers. 'The assembled crowd cheered again and again, and these demonstrations of loyalty and good feeling His Royal Highness frequently acknowledged by lifting his hat.' An even larger crowd watched as the party entered their carriages to be driven to Hopetoun House, where Miss Howard, the manageress of the Newhalls Hotel (Hawes Inn), presented the princess with a bouquet of flowers.

The town was *en fête*, with flags flying from houses on the High Street, bunting stretching across the street, a large 'Welcome' banner and a floral arch opposite the town hall. Families lined the terraced walkway above the shops, whose railings had been barricaded for safety. After the royals departed, councillors and magistrates retired to the town house to toast their success with wine and cake while the crowds thronged the streets waiting for the return of the royal party. As the afternoon wore on 'some rather noisy scenes occurred'. The streets were closed to traffic once more at 5 p.m. and the crowd became increasingly impatient as the hours passed. Finally, at 7.20 p.m., it was announced that the royals had elected to take the inland route to Dalmeny, where they were staying the night with the Roseberys and the disappointed crowds started to make their way home by cab, cycle and foot.

One of the most exotic visitors was the Shah of Persia at the end of a visit to Scotland in July 1889. *The Scotsman* opined: 'Possibly in the course of his wanderings, His Imperial Majesty had seen little, if anything, more fitted to stir his Oriental stolidity than the bridge.' By now Union Jacks were flying from the cantilevers as well as throughout the town and the Persian Royal Standard fluttered from the outward end of the Queensferry approach girder. Draped with yellow and crimson cloth, the way onto the bridge was entered through an arch above which 'His Majesty, had he been a linguist, might have read the word "Welcome"'. The workaday incline had been disguised with decorations and the jetty was carpeted in crimson. On a special stand erected beside the reception platform, hundreds of ticket holders could enjoy a grandstand view of the proceedings. The Briggers, who had been given a half-day's holiday, lined the jetty behind a barricade. Less privileged sightseers occupied every nook and cranny in the town that offered a vantage point, extra mounted police being drafted in from Midlothian to keep them in order. Their wait was unrewarded as the Shah sat impassively in his carriage making no acknowledgement of the cheering crowds. While the Shah asked questions of Sir John Fowler through an interpreter, the workmen freely commented on his jewel-encrusted tunic and his 'dusky companions'.

The crowd watched him board the *Dolphin* and swig some liquid from the spout of what looked like a gold teapot. Finally the grandeur of the bridge broke through the Shah's reserve as, cross-legged on the deck, he gestured wildly although his attendants showed no visible interest in what they saw. While standing under a crimson canopy, the sight of a cage of workmen travelling up the Inchgarvie cantilever prompted him to ask: 'How many men have been killed by falling down there?' Sir John tersely replied that the list of casualties was relatively small. Accompanied by a flotilla of pleasure craft, the *Dolphin* returned, the cheers of 200 people in the grandstand on Inchgarvie echoing across the water. The Shah slumped back into reverie, donning a fur-lined, emerald-studded cloak despite the warm July sunshine and gazing at the scene through gold spectacles attached to his hat. The citizens of Queensferry had rarely enjoyed such a curious sight.

Not all royal occasions were associated with the bridge. Three hundred people travelled by train from Queensferry to join the cheering crowds of 30,000 in the Meadows on the occasion of Queen Victoria's visit to the Edinburgh International Exhibition on 18 August 1886. Like every town and village in Scotland, the two communities celebrated Queen Victoria's Golden Jubilee on 21 June 1887. Queensferry had already raised funds for a new clock with gas-lit dials in the townhouse steeple. The publicans offered to put up 50 guineas if the five gentlemen who had petitioned the ratepayers against renewal of their hotel licences matched it; they did. William Arrol was among the subscribers and the Forth Bridge Works designed the cast-iron clock housing. On Jubilee Day, 68 bonfires burned on hilltops and headlands and the central towers of the bridge were lit by hundreds of electric arc lights indicating where work was in progress. Appropriately, Jubilee cranes were working at the top of the cantilevers, the model being named in honour of the queen. Westhofen described the towers as forming 'with their long-drawn reflections in the Firth, three pillars of fire'.

The greatest occasion of all was the opening of the bridge on 4 March 1890. The day did not start auspiciously, with a grey, hazy sky and blustery showers. In North Queensferry 'the few flags which had been erected in some half a dozen places by a few loyal burghers fluttered bravely on their slender staffs against the force of the gale'. Queensferry only boasted a single Union Jack fluttering from the flagstaff in the garden of one of the magistrates. Apart from a few hundred sightseers from Edinburgh who had caught the earliest of the special trains, the streets of Queensferry appeared much as on any working day with traders' opening their shutters and citizens going about their normal business. Only the schools had taken the public holiday, although the Briggers had no choice as a compulsory day off had been declared. Occasionally a solitary figure was spotted walking along the bridge. It seemed as if the Queensferries had become blasé about royal visitors or had become bored with their bridge.

By mid morning, however, the momentum gathered as people flooded in from Fife, West Lothian and the capital by foot, bicycle, brake, coach and rail. Detachments of police and brigades of volunteers manned the stations and the approaches to the bridge in case of trouble. Sightseers searched out the best vantage points on the steep Hawes Brae, jostled for space on the piers and hung out of the windows of the houses perched on the banks above the Forth. While waiting they admired the flags and bunting flying from HMS *Devastation*, accompanied by two gun boats and a tender. Privileged guests were allowed to view proceedings from the footpath on the viaduct approaches. Some were seen to turn back, having succumbed to vertigo and the battering of the wind which forced ladies to tie scarves and shawls round their bonnets. Some visitors sheltered in the granite arches at the end of the viaduct approach while the more adventurous continued to the place where the prince would later drive in the last of the 6.5 million rivets.

Finally three cheers went up on the Queensferry side and handkerchiefs waved as the train steamed onto the bridge and the freshening breeze whipped hats into the sea. Once the train had passed from view, some sightseers peered through the mist for a sight of the *Dolphin*, festively decked in crimson and gold, which was to take the royal party round Inchgarvie. Most people, however, decided that it was now time to enjoy themselves. They bought souvenirs of the bridge and portraits of the Prince of

Distinguished visitors to view the progress of the bridge included:

Albert 'The Good', king of Saxony, whose tour was conducted in German by Wilhelm Westhofen

Sir Henry Bessemer, whose industrial process had dramatically reduced the cost of steel

Gustave Eiffel, designer of the iconic Parisian tower

William Ewart Gladstone, British prime minister

Lord Kitchener, at the time governor of the Red Sea Territories

Leopold II, king of Belgium

Dom Pedro II, Brazil's second – and last – emperor

Lord Roberts, at the time commander-in-chief of the British army in India

The Marquis Tseng, former envoy of the emperor of China at the British Court

Wales from the stalls; they sang along with the 'wandering minstels'; they threw loose change to the 'whining beggar in stilts' or lost it to the card sharps. They overflowed onto the steps of the local eating houses. To the sound of ringing tills one shopkeeper described the event as 'a red-letter day in the history o' the Ferry'. Stablers and innkeepers were not above raising their prices for the day.

For those waiting at North Queensferry there were false hopes and excitments. Hearing the distant rumble of a train around 11 a.m., the crowd rushed towards the station. Fifteen carriages later and no sign of the royal saloon car, especially built by the Great Northern Railway for the occasion. It was in fact an unscheduled, special train carrying company directors, MPs and provosts. At 11.30 a.m. HMS *Devastation* fired a 21-gun salute and soon the royal train came into view, travelling so slowly that the few who had remained under the bridge rather than rushing to the station paid no heed to it. Twenty minutes later the crowd was finally rewarded as the Prince of Wales, in a brown overcoat, and his son, the future George V and the Duke of Edinburgh, who had travelled from Russia for the occasion, left the train, walked down the jetty and embarked the *Dolphin*, by now pitching and tossing alarmingly. The band struck up the national anthem. Half an hour later the royal party rejoined the train and headed back across the bridge. The crowd dispersed as the wind reached gale force and the heavens opened.

The Queensferry crowds had a second chance to catch a glimpse of the prince. The train stopped twice on the bridge, once for the rivet-driving ceremony and once for the prince to formally open the bridge. The prince used a silver key to operate a smaller version of one of Arrol's riveting machines. When he failed to turn the key with one hand, Arrol intervened and gave it a few vigorous twists. The prince tried again, using both hands this time, and the rivet was driven home. It was made of gold-coated, soft metal so that it could be riveted cold, the reason behind the rumour that the railway companies were too mean to splash out on solid gold. The washer round the rivet bore the inscription in blue enamel: 'The last rivet put in by His Royal Highness the Prince of Wales 4th March 1890'. The prince hastily retreated to his carriage from the wind to receive a silver gilt replica of the rivet housed in a casket made out of all the materials used in construction.

Because of the gale the latter ceremony, on a wooden platform at the end of the southern approach viaduct, had to be curtailed. Smoke and steam enveloped the royal party until an offending belching engine was removed. Clutching his hat and abandoning his prepared speech, the prince simply shouted: 'Ladies and gentlemen, I now declare the Forth Bridge open,' and retreated into his carriage with alacrity. On arrival he hastened past the cheering crowds to the warmth of the drawing shed which had been converted into a banqueting hall seating 550. The cover of the toast list featured an illustration of a train and the legend 'Thro' carriage Aberdeen to New York via Tay Bridge, *Forth Bridge*, Channel Tunnel and Alaska.'

The Scotsman considered that 'the menu for the occasion was exceedingly choice, the wines were of approved brands and the service was all that could be desired,' thanks to the catering efforts of Mr Grieve of Edinburgh's Waterloo Hotel. Dishes included 300 lark pies. Over coffee, the prince was the first to avail himself of the privilege that he granted the company – to smoke. Long speeches and toasts ensued, including the announcement of a baronetcy for Fowler and knighthoods for Baker and Arrol. Little mention was made of the contribution of the Briggers, although

Baker did attribute the success of the project as much to the individual and collective pluck of the workmen as to the scientific labours and organisation of the engineers and contractors.

The banquet in the drawing loft

Although most of the crowd started to think of home by 3 p.m. as the weather

was worsening and clouds of dust became a nuisance, royal stalwarts stood around the shed to give the prince a final send-off as he emerged from the banquet around 4.30 p.m. Thereafter 'inebriates made their presence known' and townspeople settled beside their fires to contemplate Queensferry without the Briggers.

CHAPTER 6
DAYS, YEARS AND DECADES

What Benjamin Baker described as 'a romantic chapter from a fairy-tale of science' was over. With the departure of the Briggers, the Queensferries were like ghost towns. Streets and inns were no longer thronged with workmen and the estuary seemed strangely empty without the work boats plying to and fro. A large number of houses stood empty and landlords were forced to think the unthinkable, of reducing rents which they had doubled in recent years. Their profits had already been eroded by the 'leaving of keys under doors and moonlight flittings'. Shopkeepers had time on their hands and unsold goods on their shelves. According to the *Queensferry Observer*, 'Nothing, perhaps, so truly indicates the return to the old system of things than the record of the police courts, where now the appearance of a prisoner seems to be rather the exception than the rule.'

The future of the two communities had been a matter for debate and speculation long before the first trains crossed the bridge. As early as July 1888 the *Queensferry Observer* posed the questions: 'In what state will Queensferry be left?'; 'How can new markets be created to sustain the improved facilities?' The editor's solution was to promote the town as a seaside resort with yachting basins and a promenade, the latter finally being built in the 1930s. Presciently, he also suggested that the town 'would ultimately become one of the attractive suburbs of the Scottish capital'. There was even a suggestion that the drill roads could be converted into tennis courts. Less ambitiously the contractor's ambulance cart could be purchased by the town and now that it had achieved a decent water supply, a fire station should become a priority. In practice the town gained confidence from the bridge and the anticipated number of visitors. Plans which had been debated for years were dusted down and moved forward. These included not only a much-needed new cemetery but also a grand civic hall for concerts, entertainments and meetings, which had won support from the Earl of Rosebery.

Step by step, the giant undertaking was dismantled. The North British had hoped to retain the sheds at Queensferry but Lord Rosebery did not wish his town to be spoilt by industry. Soon turnips sprouted in the fields which had so recently been a sea of mud and human activity, and cranes vanished from the jetties, themselves

Before very long, little will be left to show that such large works ever existed.

Queensferry Observer, 1890

Contemporary view from east of the bridge

121

The ghost town

scheduled for demolition. Four months later the engineering plant went under the auctioneer's hammer attracting 2,700 prospective buyers, some of whom had travelled from the Continent. Bidding was so brisk that the auctioneer and his son had to dash from one lot to the next, entertaining the crowd with their dry humour as they went. HMS *Monarch* lifted the telegraph cables which had for decades proved a hazard to shipping in the Forth: they now crossed the bridge in special tubes beside the railway line. There was even talk of using the bridge to bring the telephone to Queensferry. By the end of August, laden goods trains had removed most of the machinery and timber and the canteens had been demolished although the Queensferry huts still blotted the landscape. As late as spring 1891 the town rocked as the concrete foundations of machines and huts were finally blown up.

Little remained to remind residents of the vast undertaking except the occasional rumble of a train and the ever-present monument of the bridge itself. After a few weeks they became accustomed to the noise of the trains, which averaged 140 a day, and could forecast the weather by the different way the sound carried depending on the strength and direction of the wind. The presence of the trains, however, drove home the one thing that the two communities had overlooked, namely that the bridge would result in the closure of their local railway stations. The station in Queensferry town centre on the line to the ferry at Port Edgar closed two days after the bridge opened. Residents were now faced with a mile's walk to Dalmeny station to catch a train. North Queensferry residents too were up in arms, expressing 'surprise and indignation' at a public meeting after they lost their station close to the ferry landing. Its closure meant a two-mile walk to Inverkeithing to board the trains that people could see passing above their heads. Four months later their protests resulted in a new station at North Queensferry, served by a dozen trains on weekdays and four on Sundays. Even after the bridge opened, ferryman John Arthur continued to run the traditional Queen's passage between the two communities, purchasing the Forth Bridge work boat *Dolphin* to replace his ageing *John Beaumont*. One casualty of the bridge was the local newspaper. Founded when construction activity was at its height, by 1891 the *Queensferry Observer* no longer had a viable market nor enough news to fill its pages.

The opening of the bridge had an impact on more distant communities. While Dunfermline enjoyed the novelty of a Sunday train service, Burntisland lost out when over 150 men lost their jobs. This was occasioned by the closure of the cargo ferry to Granton and the distribution of work, from engine driving to carriage cleaning, more widely through the new rail network. By spring 1890 houses were proving difficult to let or sell. Newspapers arrived an hour later than before and the post was delayed by at least 40 minutes, travelling first to Kirkcaldy and then back to Burntisland by a slow train. Inverkeithing suffered too, its streets assuming 'the quiet Sunday appearance which they exhibited before the great Bridge was begun'.

Of Queensferry, Edinburgh's *Evening News* gloomily predicted: 'The ancient burgh, which has been roused from its previous lethargy by the presence in its near vicinity of the great structure, with the necessary accompaniment of its army of workmen and legion of visitors, may be fated to return to comparative oblivion with the completion of the work.'

It was not to be. The new bridge became the eighth wonder of the Victorian world. It was the first great project to capture the nation's imagination since the Crystal Palace. Back in 1884 a Leeds newspaper had described the building of the bridge as a task 'as impossible as the construction of the Tower of Babel, a triumph of engineering skill to eclipse the Ship Canal which has turned Africa into an island and a work which will reduce the pyramids to mere child's play and, in all likelihood, lead to a revolution in the art of constructing bridges of this description'. By 1890 the impossible had been achieved. Those unable to visit in person celebrated the bridge's success in the same way they had eagerly followed its construction: through newspapers and illustrated magazines. They bought mementoes, from trinket boxes and models to china ornaments and embroidered pictures. One of the first to cash in was Queensferry entrepreneur Messrs Francis Rae & Co. who produced an album of views of the bridge and other local attractions, 'as good a shilling's worth as is going in the souvenir line'.

The verdict on the bridge was largely favourable. Fresh from completing his own iconic structure, Gustave Eiffel considered it 'the greatest wonder of the century'. The views of local people are largely unrecorded although one resident was moved to verse.

How little thought we years ago
As by thy shores we wandered on,
That man would so thy passion stem,
Thy billows chaff, nor heed thy moan –
Thou Forth!

For now with genius man hath cast
A mighty iron arm to bind
Thy shores, which groan beneath a strength
That bridges all that storms can mind –
Thou Forth!

Not everyone, however, liked the bridge, art critics especially being divided in their reaction. Architect Alfred Waterhouse wrote to Fowler: 'One feature particularly delights me – the absence of all ornament. Any architectural detail borrowed from any style would have been out of place in such a work.' William Morris, the Arts and Crafts designer, took an opposing view, describing the bridge as 'that supreme expression of all ugliness'. Baker riposted that Morris had not the faintest knowledge of the function that the bridge had to serve.

The London correspondent of the *New York Tribune* was among the most outspoken. He demanded that 'The contractor, designer, engineer of the Forth Bridge ought, each and all, to be hanged from the topmost angle of its cantilevers. That is the only thing that would improve the appearance of this hideous structure, except dynamite. It totally ruins some of the most beautiful scenery in Scotland or the world . . . monstrous, horrible, huge, shapeless and ugly beyond the ugliest dreams of Dante. A nightmare in granite and iron . . . It is simply the ugliest thing in the world.'

Even Westhofen conceded that 'this bridge or any other bridge, must be a discordant feature in a pastoral landscape'. He had often stood in Dalmeny Park overlooking the Forth with the backdrop of the Ochil hills and the distant outlines of the Trossachs and Ben Lomond 'a view hard to be excelled in any part of the world'. The bridge was also criticised as expensive and over-engineered. One such critic lived to regret his words. Theodore Cooper's design for the Quebec Bridge over Canada's St Lawrence River used much lighter structural members. It collapsed twice while under construction, killing 87 workers.

After the bridge opened tourists replaced workmen on Queensferry High Street, the town being packed on Good Friday and the Edinburgh spring holiday. In August 1890 it was estimated that 2,000 visitors a day came by coach, sea and rail, resulting in serious delays and inconvenience at Edinburgh's Waverley Station. Coaches, each carrying 29 passengers, made 52 runs daily to view the bridge while passengers on the five steamer excursions run by Galloway's sailed right under the bridge itself. Galloway's and the North British finally settled their differences so that day-trippers could buy a combined ticket to sail under the bridge and return over it by rail. One enterprising Queensferry fisherman found himself in prison when he used the pier illegally as a starting point for his bridge boat trips. Local hotels enjoyed their busiest summer ever.

When the first novelty wore off and long winter nights gave time for reflection, some residents missed the Briggers, or at least their money: 'Since the opening of the bridge Queensferry has become more dead than alive than ever.' Port Edgar too was suffering although there was a hint of things to come when Admiral Fairfax visited to assess its suitability as a base for a naval training ship. By spring 1891 business was looking up again when over a thousand people poured into the town during the Easter holiday. The May holiday too was busy with laden four-in-hands leaving

Wish you were here

Waverley Steps every few minutes for Queensferry. The summer of 1891 proved that the bridge was not a nine-day wonder but a permanent tourist attraction.

The bridge proved a popular destination for cyclists, indulging in an activity which became a craze of epidemic proportions in the 1890s. Although warning signs had been posted at the top of the steep Hawes Brae, some cyclists simply could not stop. A day out for two young men ended when their bicycles hit a wall at the bottom of the brae and they had to hire a cab back to the capital.

Those hoping to stroll along the bridge's footpath were largely thwarted. Women were disallowed entirely on safety grounds and men had to seek permission from the bridge manager, whose agreement was 'very difficult to procure'.

Presumably from personal experience, the editor of the *Queensferry Observer* was of the view that 'in the majority of people, if they were allowed to pass along the bridge on foot, and experience the sensation of a train flying past them, they would not ask to go back a second time . . . When a train is passing the pedestrian has to cling on to the rail of the bridge with only something like two feet between him and the flying mass.'

Drink remained a problem even with the Briggers gone. In April 1890 ten new licences for the sale of drink were approved and 16 renewed including the reinstatement of the status of the four hotels. Later that year the first temperance hotel opened its liquorless doors, a novelty for Queensferry, which came second only to the Borders market town of Jedburgh in terms of the number of licences per 1,000 residents. By the same measure in 1899 the town topped the Scottish league table of convictions for drunkenness. One correspondent writing in the *West Lothian Courier* blamed the day-trippers, especially those who arrived on Sundays, claiming that only a sixth of convictions could be attributed to local people.

The stream of distinguished visitors continued – the Lord Mayor of London in June 1890, Gladstone in October, and Queen Victoria regularly crossed on her way to her Highland holiday retreat at Balmoral. In 1896 the Chinese ambassador Ling Hung Chang fired endless questions on everything from heat expansion to salaries while inscrutably giving nothing away about his own impressions of the bridge. No longer responsible for answering such barrages of questions, Fowler, now in his seventies, retired to enjoy life in Bournemouth and on his West Highland estate of Braemore. Other designers and contractors filed their Forth Bridge drawings and marketed their newly gained experience. Baker designed the Aswan Dam on the Nile before succumbing to a fatal fainting attack in 1907. Sir William Arrol, of whom it was said, 'He made the Forth Bridge, and the Forth Bridge made him,' went on to build another of the world's iconic structures, London's Tower Bridge, as well as entering politics and marrying twice more. He died of pneumonia in 1913, a much-honoured man. On the completion of the bridge, resident engineer Andrew Biggart was offered a partnership in Arrol's firm, where he rose to be chairman. He called his mansion in Glasgow's Pollokshields 'Inchgarvie'. Kaichi Watanabe, the young engineer who so vividly demonstrated the principle of the cantilever, returned to Japan in 1888, working his way to the top in the railway, engineering, shipbuilding and energy industries and pioneering the introduction of electric railways.

The Fowler family at their Braemore estate

One engineer left a friend behind under 'Lost and Found', The Scotsman of Friday 29 May 1891 carried the following poignant advertisement: 'Manchester Terrier (black), with a few tan spots, split ear, wears chain collar with owner's name on it, and answers to "Brer". Lost between Queensferry and Cramond. Finder communicate with Westhofen, South Queensferry.' By this time Westhofen had emigrated to Cape Town, where he combined his engineering practice with illustrating books about South Africa. The breed was known for its devoted loyalty and Brer must have gone off to look for his master while being taken for a walk along the Dalmeny foreshore.

While the bridge made or enhanced the careers of senior management, little is known of what happened to the Briggers. Many would have found work on other construction or shipbuilding projects throughout Britain. Some stayed in the locality where they had put down roots, seeking work as labourers in the fields and mines.

'A few finding the British Isles of too little scope are winging their flight across the herring pond to the land of liberty, the land of the stars and stripes.' The *Queensferry Observer* attributed this exodus partly to the influence of Mr McKay, the foreman of the joiner's shop, whose son had settled in 1888 in a 'fashionable resort known as Stratton [sic] Island' near New York. Mr McKay must have regaled workmates with glowing reports of his son's new life.

One joiner who followed his example was Willie Gilmour. 'Last Thursday evening a select number of friends met to bid him God-speed . . . During his stay in our midst, he has made many friends by his sociable qualities. In the course of the evening, he was presented with a very appropriate present in the form of a handsome travelling bag . . . A happy hour was afterwards spent in song and sentiment.'

When the bridge opened, only the parts visible to the public had been painted. It took a further year to complete the job. Working from their moveable stages painters thrust their brushes into every inaccessible nook and cranny where rust could take hold. The insides of the cantilever tubes were painted white, for the first and only time, so that the slightest leak could be detected by the trail of water it left.

Barely had the bridge's paint dried before the endless task of repainting its 145 acres of surfaces began. A painting workshop was built on the north side of the river as the base for the workmen, known as ship riggers, who spent their days scrambling up and down the staging, brush in hand. The first recorded accident among the ship riggers had occurred during the initial painting in 1889 when a former plater, Thomas McLeod, who had been employed on the bridge since the start, was hit on the head by a falling plank while coating the bracing above the Inchgarvie viaduct. He fractured his skull and took Arrol to court for damages. George Mathieson was the first ship rigger to fall victim to the bridge only three days after its opening, dying in the ambulance house two and a half hours after he fell and fractured his skull.

The men risked not only falls but also painter's colic from breathing lead paint fumes. On 7 September 1891, 45-year-old John Whigham complained to workmates of having stomach cramps. Although he was missed after dinner no one was seriously concerned until the watchman reported that he had not handed in his pass and his flask was found. As he had to climb over the handrail and down a few steps on a ladder to reach the area he was painting, the fear was that, gripped by a sudden spasm of painter's colic, he had fallen off the bridge. The official record simply read: 'Drowning. Body not recovered.'

On 29 September 1891 one of the bridge's worst accidents occurred when a temporary scaffold on the south viaduct collapsed in a gusting gale, killing 50-year-old father-of-eight James Smith and riveter James Jack. After struggling for some time to maintain his grip and screaming for help, a third victim Robert Hughes plummeted 120 feet, striking a grassy slope 'with a sickening thud' in front of a crowd of horrified onlookers. The men had built the scaffolding themselves. Teenage rivet heater William McLucas, whose mother was already a Brigger's widow, was lucky. He was kneeling on one of the end frames which was attached to the tackling as well as to the plank that had given way.

Trains were a new hazard. Only a week after the official opening a goods train derailed when a gale whipped a tarpaulin off a wagon and three bales of esparto grass and a wooden case were blown onto the line. The wheels of a wagon caught on the case and became derailed. But instead of the wagon careering out of control, its wheels continued to run so smoothly in the adjacent safety trough that the train had reached the approach viaduct before anyone was aware of a problem. Had the goods fallen onto the other line the consequences might have been very different.

A month later tragedy struck when children were larking about on the workmen's platform near the Forth Bridge Works. One six year old was lying on the platform with his head hanging over the edge when a goods train approached. Because of the curve in the line the driver did not spot the boy in time. The step of the guard's van hit the child on the side of his head. Fog also reduced visibility for

A French meat-extract company adopted the bridge as an icon

Ponts remarquables.

Pont sur le Forth à Queensferry (Ecosse).

PEPTONE DE VIANDE DE LA COMPAGNIE LIEBIG.

Voir au verso.

train drivers. In February 1891 two men were painting the southern approach viaduct girders when a train ran them down. Both were severely injured.

Coat upon coat upon coat – over the ensuing century 'painting the Forth Bridge' became a byword for any never-ending task. Its proverbial status showed how much the bridge had won its place in people's affections throughout the world. It appeared on banknotes and stamps and on breakfast tables, from Alberta to Zanzibar, as well as in the form of postcards from those lucky enough to see it for themselves. The iconic structure earned its place in literature through John Buchan's *The Thirty-Nine Steps* and more recently Iain Banks' *The Bridge*. It has featured in computer games and television documentaries as well as on celluloid, most notably in Alfred Hitchcock's 1935 classic film of Buchan's novel when hero Richard Hannay escaped his pursuers by jumping from a train as it crossed the bridge, albeit a studio mock-up.

The bridge has seen its share of drama. In the run-up to the First World War

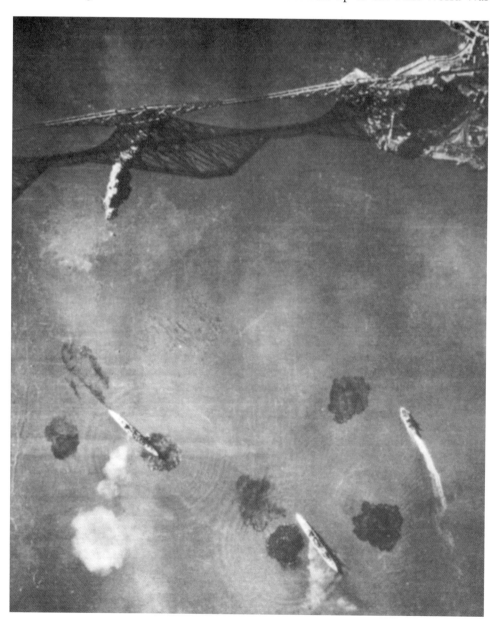

View of the bridge from a Luftwaffe plane

Inchgarvie was fortified with gun emplacements in case the enemy sailed up the Forth, and from 1917 Port Edgar became a training base for destroyers. On 16 October 1939, during the UK's first bombing raid, German planes flew over the Forth, their target the battleships anchored below. One passenger, watching from a train as it entered the first arch at the southern end of the bridge, recalled seeing a giant water-spout as high as the bridge as HMS *Southampton* took a direct hit. The Germans regarded destroying such a famous landmark as a propaganda coup. There is a legend that in desperation, they took an out-of-focus picture of the bridge and printed a negative image in German newspapers, claiming that the outline of Inchgarvie was dust and smoke billowing up after the bridge had been bombed. After the end of the war, the increasing volume of motor traffic led to plans for a road bridge, finally realised in 1964, one of which involved running a road way above the rail track on the bridge.

'The Last Brigger', introduced to a Dunfermline art class as the last man alive who had worked on the bridge, painted by young artist Sandy Craig in the 1950s

Anniversaries of the bridge's opening were celebrated, especially its centenary in 1990. A giant clock, mounted on the central span, counted the days before the new millennium and in 2005 the bridge's lights blushed red for the charity initiative Comic Relief. Between-times routine maintenance continued, although it was all too tempting to cut back when budgets were tight. By the 1990s residents of North Queensferry complained of lumps of rust raining on their houses and cars and alarmist headlines suggested that the bridge would have to be closed for good. Although the old lady's makeup was beginning to crack, on inspection her body was found to be remarkably sound with only 100 of her 54,000 tons of steelwork identified as needing replaced.

At the time of writing, in the largest repair programme of the bridge's history, workers are stripping off layers of paint until a sound surface is achieved – shot-blasting back to bare metal is not possible as corrosion would start before a large enough area could be exposed for painting – before applying an undercoat and two coats of glass-flake epoxy resin. Already used on North Sea oil rigs, the resin is so tough that painting the Forth Bridge may become a thing of the past for at least a generation to come.

Its future assured, the bridge is likely to span the Forth for as long as there are trains to cross it. It is much more than a bridge: it is an icon recognised the world over, a byword for strength and a triumphant memorial to Victorian engineering.

What has been largely forgotten, until now, is the energy and skill, the hopes and frustrations, the blood and broken bones of the thousands of men who created this icon. They came, they built and then they left again. We can only catch glimpses of who they were, how they lived and the fate of the 73 or more who died for the bridge. For the first time we can reveal some of their faces, appearing from the ghostly shadows, the dots and pinpricks of contemporary photographs. What they thought and how they felt as they hammered in a rivet, passed through the caisson's airlock or waited for a work boat at the end of a shift, we can only imagine.

The story does not end there. Even as this book goes to press, more Briggers are coming to light. David Carmichael lived in North Queensferry and found work on the bridge as a holder-on. After its opening he joined the maintenance team, retiring in 1933. Fellow workers presented him with a mantlepiece clock to mark his 40 years of service. He then opened a confectionary kiosk on North Queensferry pier enter-taining visitors with photographs and stories of 'his bridge'.

David Carmichael with grandson Ian Wall

APPENDICES

APPENDIX 1: Briggers who died as a direct result of working on the Forth Bridge

Name	Date of death	Occupation	Age at death	How died
Robert Adamson	14 July 1884	General labourer	37	Crushed by wagons
John Aitken	13 September 1889	Iron riveter's labourer	17	Fell from the bridge
John Ashwood	25 October 1888	Bridge labourer	28	Fell from the bridge
Michael Bearregan	9 May 1887	Rivet heater	18	Fell from the bridge
Thomas Birrell	7 April 1887	Labourer	59	Fell from the bridge
William Brown	6 August 1888	Bridge labourer	22	Fell from the bridge
John Burnett	6 August 1890	Pattern maker (Journeyman)	25	Fell during initial painting
James Christie	27 September 1886	Seaman	34	Fell from rigging of Forth Bridge steamer Beamer
David Clark	13 September 1888	Rivet catcher	13	Fell from the bridge
Thomas Conroy	18 December 1889	Railway miner	23	Fell from scaffold
Joseph Courtney	28 November 1887	General labourer	21	Fell from the bridge
John Curran	31 July 1888	Bridge labourer	23	Fell from the bridge
James Davidson	15 November 1886	Carpenter's labourer	29	Fell from the bridge
Edward Davies	9 September 1887	Joiner's labourer	23	Struck by drift
Alexander Douglas	9 August 1886	Bridge labourer	45	Rupture of liver and spleen
John Fairless	30 June 1884	Steamship captain	46	Struck by falling stone
William Fairley	16 July 1888	Boatman	48	Drowned after boating accident
Alexander Ferguson	25 March 1885	Carpenter	31	Drowned in caisson
George Fowler	2 January 1892	Railway lorryman	34	Caisson disease
Thomas Fraser	30 October 1888	Gatekeeper	61	Found drowned on beach
John Gallacher	28 July 1887	Hand driller	21	Fell from the bridge
John Gibson	25 March 1885	Boiler maker	35	Drowned in caisson
Thomas Harris	27 November 1883	Storekeeper	16	Drowned at the jetty

Name	Date of death	Occupation	Age at death	How died
George Hendry	2 June 1887	Riveter	32	Fell from the bridge
James Hennan	13 September 1886	Rivet Heater	23	Fell from the bridge
William Howie	6 March 1889	Labourer – railway approaches	60	Crushed by stone
Francis James Watt	17 January 1890	Rigger	22	Fell from the bridge
Robert Hughes	29 September 1891	Riveter	28	Fell from the bridge
James Campbell Jack	29 September 1891	Boiler maker	30	Fell from the bridge
William Jamieson	30 August 1889	Rigger	45	Fell from the bridge
Robert Latto	21 November 1887	Machineman	21	Crushed by machinery
Nicolia Liberale	11 March 1885	Engine man	40	No cause given
James MacCallum	23 March 1888	Carpenter	23	Fell from the bridge
John McClinchie	28 July 1887	Hand driller	34	Fell from the bridge
John McDade	16 December 1885	Bridge labourer	32	Stuck by steel plate
Kenneth McDonald	26 November 1884	Fitter	41	Fall from scaffold
Patrick McGarry	18 August 1887	General labourer	16	Fell from the bridge
Patrick McGinn	17 September 1889	Bridge labourer	18	Struck by timber
James McLachlan	8 October 1888	Railway labourer	35	Hut fire
Frank McLean	17 October 1887	General labourer	23	Fell from the bridge
Peter McLucas	3 December 1887	Road labourer	61	Fell from the bridge
Montague McNab	1 November 1887	Holder-on	37	Fractured skull
Peter McCowat	13 September 1887	Plater	49	Struck by falling timber
Terence Martin	8 October 1888	Railway labourer	50	Hut fire
George Mathieson	17 March 1890	Painter (journeyman)	33	Fell from bridge
John Millar	15 May 1888	Carter on rail approaches	41	Crushed by wagons
James Munday	11 October 1887	Plater	20	Fell from the bridge
Alexander Ogilvie	27 May 1886	Contractor's carter	26	Crushed by wagon
John O'Neil	3 August 1887	Bridge labourer	16	Fell from the bridge
Angus Paterson	6 October 1886	Mechanical engineer's labourer	21	Crushed by plate
William Perkins	27 May 1887	Rigger	41	Fractured skull
Robert Reid	18 March 1888	Night watchman	32	Drowned
William Rendill	1 November 1887	Carpenter	30	Fractured neck
Owen Rennie	17 June 1889	Rigger	28	Fell from the bridge
Hugh Richardson	24 September 1889	Plater at Forth Bridge Works	25	Struck by timber
Thomas Roberts	16 July 1888	Boatman	32	Drowned after boat accident
Joseph Robertson	24 February 1885	Labourer	42	Struck by plank
Michael Rogers	16 December 1889	Labourer on rail approaches	21	Crushed
Thomas Shannon	9 December 1887	Rivet -catcher	14	Fell from the bridge
Hume Smith	19 June 1888	Mechanical engineer's labourer	36	Fractured skull
James Smith	29 September 1891	Rigger	49	Fell from the bridge
Alexander Steel	6 May 1887	Iron worker's labourer	26	Fell from the bridge
James Syme	3 August 1887	Painter	53	Fell from the bridge
David Thomson	13 February 1889	Stone miner on rail approaches	32	Crushed by stone
Robert Thomson	27 October 1888	Mechanical engineer's labourer	38	Fractured skull
Richard Timney	16 September 1886	Engineman	17	Fell from the bridge

Name	Date of death	Occupation	Age at death	How died
Thomas Vennert	22 April 1889	Joiner's labourer	35	Crushed by machinery
John Ward	8 October 1888	Railway labourer	40	Hut fire
James Wark	3 October 1888	Labourer	26	Drowned
Eugen Weise	29 March 1887	Rigger	20	Fell from the bridge
Martin Welsh	2 June 1887	Riveter	25	Fell from the bridge
John Whigham	7 September 1891	Painter	46	Drowned – after suspected painter's colic
William Wright	25 August 1888	Labourer	19	Fell from the bridge

APPENDIX 2: Other Forth Bridge-related deaths

Died of natural causes during construction

Name	Date of death	Occupation	Age at death	How died
Egatz Arcangelo Di Bortolo	6 September 1885	Mechanical engineer	43	Probably typhoid
Charles Dickson	14 November 1886	Plater	42	Bleeding lungs
William Forrest	13 January 1885	Joiner	60	Heart disease
Patrick Gobbins	3 June 1889	Bridge labourer	45	Tuberculosis
John Hart	6 September 1886	Plater's labourer	24	Heart disease
Lachlan McDonald	9 April 1889	Bridge labourer	19	Cardiac valvular
Charles Petterson	30 March 1885	Bridge labourer	27	Meningitis
David Scott	18 October 1885	Inspector	53	Stroke
Frederick Scott	17 March 1885	Bridge labourer	48	Bright's disease
Alexander Smart	27 November 1884	Submarine diver	37	Heart condition
David Wilson	21 September 1889	General labourer	30	Pulmonary phthisis
Robert Wilson	28 February 1889	Storekeeper	67	Cerebral haemorrhage

Unconfirmed bridge deaths

Name	Date of death	Occupation	Age at death	How died
John Conway	15–17 October 1887	General labourer	22	Possible foul play
Colin McArthur	26 September 1886	Pilot and harbour-master	52	Drowned in boating accident
Robert Miller	13 March 1889	Time-keeper	33	Fell from Forth Bridge steamer *Tay Bridge*
George Reid	5 July 1887	Rope maker	48	Died drinking in Edinburgh

Other deaths possibly linked to the bridge

Name	Date of death	Occupation	Age at death	How died
Richard Harvey	31 August 1891	Seaman, Royal Navy	31	Navy death
James Jack	10 January 1891	Lighthouse keeper	62	Bridge death after completion
Hugh McDonald	26 November 1889	Railway navvy	about 34	Crushed by steam navvy
Richard Morton	25 July 1884	Rigger/diver	38	Drowned on naval work
Thomas Wilson	22 April 1892	Painter	52	Death after completion
George Wright	22 April 1892	Painter	44	Death after completion

Outside scope of research

Name	Date of death	Occupation	Age at death	How died
Patrick Campbell	14 August 1888	Railway labourer	50	Fell in front of a train
William Cunningham	1 June 1888	Stoker	42	Accident on steamer
Davidson	13 February 1888	Not known	Not known	Not known
Frederick Duff	17 December 1889	Mason	28	Railway death
William Greig	6 November 1888	Labourer	35	Jumped out of train
Alfred Hall	16 January 1890	Railway miner	24	Fractured skull
Daniel Keenan	9 December 1885	Railway surfaceman	45	Railway accident
Patrick McGinty	9 October 1887	Bridge labourer	26	Died suddenly
John Murray	11 March 1890	Labourer	25	Not known
Thomas Quinn	15 October 1888	Railway labourer	40	Effusion of blood on the brain
William Robertson	4 January 1889	Mason	51	Died in poorhouse from fractured skull

SOURCES

Records

Criminal Records (JP Court) 1887–90, City of Edinburgh Council Archives

Police Commissioners' Reports 1882–92, City of Edinburgh Council Archives

Record of Criminal Trials (Northern Division) 1883–6, City of Edinburgh Council Archives

Records of Casualties 1885–90, City of Edinburgh Council Archives

Register of West Lothian Constabulary 1885–1921, City of Edinburgh Council Archives

Royal Infirmary of Edinburgh Registers 1883–7, Lothian Health Services Archive, Edinburgh University Library

Statutory Deaths Index 1884–90, New Register House (General Register Office)

Town Council Minutes, Queensferry 1883–90 ,City of Edinburgh Council Archives

Newspapers and Journals

Dunfermline Journal
Edinburgh Evening News
Graphic
Illustrated London News
The New York Times
Queensferry Observer
Pall Mall Gazette
The Scotsman
West Lothian Courier

Other contemporary sources

Baker, Sir B., *Bridging the Firth of Forth. Lecture delivered at the Royal Institution, Friday, May 20th, 1887* (London, Bedford Press, 1887)

Biggart, A., *The Forth Bridge Works. 55th Report of the British Association for the Advancement of Science held in Aberdeen, 1885* (London, BAAS, 1885)

Fyfe, W., *Summer Life on Land and Water at South Queensferry* (Edinburgh, 1851)

Groome, F. (ed.), *The Ordnance Gazetteer of Scotland: A Survey of Scottish Topography, Statistical, Biographical and Historical* (Edinburgh, Thomas C. Jack, 1882–85)

Hunter, Dr J., 'Compressed Air – Its Physiological and Pathological Effects' (Edinburgh, unpublished thesis, 1887.)

Orrock, T., *Fortha's Lyrics and Other Poems with a Descriptive Account of South Queensferry and Its Surroundings* (Edinburgh, 1880)

Phillips, P., *The Forth Bridge In Its Various Stages Of Construction And Compared With The Most Notable Bridges Of The World* (Edinburgh, R. Grant & Sons, 1890)

Westhofen, W., 'The Forth Bridge' in *Engineering Magazine* (London, 1890)

ACKNOWLEDGEMENTS

We should like to thank the many organisations and individuals who assisted the research team including: BRB (Residuary) Ltd; City of Edinburgh Council Archives; Dunfermline Carnegie Library; Michael Gray, independent curator and photographer; Dr Adam Hay; The Historical Diving Society (John Bevan and Gary Wallace-Potter); Lothian Health Services Archive; The National Archives of Scotland; The National Library of Scotland; North Queensferry Heritage Trust; Queensferry History Group; Dr John Ross, Aberdeen University; Scottish National Portrait Gallery (Sara Stevenson); Val Williams.

Particular thanks to: Chitaru Asahina; Mrs J. Cumming; Alan Dowsett; Sheila Hay, genealogist; Jim Magnuson; Alan Meldrum, technical support; Gordon Muir, graphic designer; Alice Seymour; the family of the late Norman Steven; Ian Stewart, genealogist; Ian Wall; Elspeth Wills, writer.

Picture acknowledgements

National Archive of Scotland photographs by permission of BRB (Residuary) Ltd: Front cover NAS BR/FOR/4/34/25; ii NAS BR/FOR/4/34/429; iv–v NAS BR/FOR/4/34/ 183 (detail); vi NAS BR/FOR/4/34/ 88 (detail); viii NAS BR/FOR/4/34/250 (detail); 11 NAS BR/FOR/4/34/ 204 (detail); 12 NAS BR/FOR/4/34/233; 14–15 NAS BR/FOR/4/34/ 110 (detail); 16 (NAS BR/FOR/4/34/174 (detail); 17 NAS BR/FOR/4/34/45; 18 NAS BR/FOR/4/34/334 (detail); 20 NAS BR/FOR/4/34/397; 21 (top) NAS BR/FOR/4/34/417 (detail); 21 (bottom) NAS BR/FOR/4/34/118 (detail); 22 NAS BR/FOR/4/34/73 (detail); 23 NAS BR/FOR/4/34/448 (detail); 24 NAS BR/FOR/4/34/89; 25 NAS BR/FOR/4/34/37; 26 NAS BR/FOR/4/34/66; 28 NAS BR/FOR/4/34/ 318 (detail); 30 NAS BR/FOR/4/34/352 (detail); 31 NAS BR/FOR/4/34/420; 32 (top) NAS BR/FOR/4/34/25 (detail); 32 (bottom) NAS BR/FOR/4/34/86; 33 NAS BR/FOR/4/34/88 (detail); 34 NAS BR/FOR/4/34/116; 35 NAS BR/FOR/4/34/167 (detail); 38 NAS BR/FOR/4/34/424a; 39 NAS BR/FOR/4/34/51 (detail); 40–41 NAS BR/FOR/4/34/172; 44–45 NAS BR/FOR/4/34/415; 47 NAS BR/FOR/4/34/138; 48 (top) NAS BR/FOR/4/34/13 (detail);

48 (bottom) NAS BR/FOR/4/34/15 (detail); 49 NAS BR/FOR/4/34/13 (detail); 50 (top) NAS BR/FOR/4/34/347 (detail); 50 (bottom) NAS BR/FOR/4/34/347 (detail); 51 NAS BR/FOR/4/34/51 (detail); 52 NAS BR/FOR/4/34/203 (detail); 53 NAS BR/FOR/4/34/2; 54 NAS BR/FOR/4/34/236 (detail); 56 NAS BR/FOR/4/34/315; 58 NAS BR/FOR/4/34/302 (detail); 60 NAS BR/FOR/4/34/334 (detail); 61 NAS BR/FOR/4/34/54 (detail); 62–63 NAS BR/FOR/4/34/26; 64 NAS BR/FOR/4/34/25 (detail); 65 NAS BR/FOR/4/34/66 (detail); 66 NAS BR/FOR/4/34/184 (detail); 67 NAS BR/FOR/4/34/138 (detail); 69 (top) NAS BR/FOR/4/34/176 (detail); 69 (bottom) NAS BR/FOR/4/34/176 (detail); 71 NAS BR/FOR/4/34/245 (detail); 72 NAS BR/FOR/4/34/47 (detail); 73 (top) NAS BR/FOR/4/34/318 (detail); 73 (bottom) NAS BR/FOR/4/34/250 (detail); 74 NAS BR/FOR/4/34/27 (detail); 75 NAS BR/FOR/4/34/271 (detail); 76 NAS BR/FOR/4/34/47 (detail); 77 NAS BR/FOR/4/34/297 (detail); 78 NAS BR/FOR/4/34/159; 81 (left) NAS BR/FOR/4/34/426 (detail); 81 (right) NAS BR/FOR/4/34/275 (detail); 83 NAS BR/FOR/4/34/306 (detail); 84 NAS BR/FOR/4/34/25 (detail); 85 NAS BR/FOR/4/34/54 (detail); 87 (top) NAS BR/FOR/4/34/1b (detail); 87 (bottom) NAS BR/FOR/4/34/471 (detail); 89 (top) NAS BR/FOR/4/34/471 (detail); 89 (bottom) NAS BR/FOR/4/34/471 (detail); 90 NAS BR/FOR/4/34/13 (detail); 91 (top) NAS BR/FOR/4/34/141 (detail); 91 (bottom) NAS BR/FOR/4/34/167 (detail); 92 (NAS BR/FOR/4/34/365 (detail); 93 NAS BR/FOR/4/34/250 (detail); 95 NAS BR/FOR/4/34/17 (detail); 96 NAS BR/FOR/4/34/164; 98 NAS BR/FOR/4/34/21 (detail); 100–101 NAS BR/FOR/4/34/58; 102 NASBR/FOR/4/34/19 (detail); 104–105 NAS BR/FOR/4/34/223; 107 NAS BR/FOR/4/34/29 (detail); 112 NAS BR/FOR/4/34/262; 113 NAS BR/FOR/4/34/99 (detail); 127 NAS BR/FOR/4/34/230; 132 NAS BR/FOR/4/34/21 (detail); *The Forth Bridge*, Philip Phillips: 27; *The Forth Bridge*, Wilhelm Westhofen: i, x, 5 (bottom), 8 (top) and 8 (bottom); *Illustrated London News*: 4–5, 116 and 118–119; *The Graphic*: 67; *Scientific American*: 120; *The War Illustrated*: 130; by courtesy of Jim Magnuson: 86; by courtesy of the family of Norman Steven: 131(top); by courtesy of Ian Wall: 131 (bottom); by courtesy of Alan Dowsett, from *Die Forth-Brücke*, G. Barkhausen, Berlin, 1889: 13; research team's collection of contemporary engravings, postcards and photographs: 2, 4 (bottom), 10, 55, 106, 109, 122, 124 top, 124 (middle), 124 (bottom), 126 and 129.

INDEX